P9-CQL-230

Copper and Zambia

The Wharton Econometric Studies Series

Wharton Econometric Forecasting Associates and
Economics Research Unit
The University of Pennsylvania
F. Gerard Adams and Lawrence R. Klein, Coordinators

Stabilizing World Commodity Markets
Edited by F. Gerard Adams and Sonia A. Klein
Econometric Modeling of World Commodity Policy
Edited by F. Gerard Adams and Jere Behrman
An Introduction to Econometric Forecasting and Forecasting Models
Lawrence R. Klein and Richard M. Young
Modeling the Multiregional Economic System
Edited by F. Gerard Adams and Norman J. Glickman
Coffee and the Ivory Coast: An Econometric Study
Theophilos Priovolos
The Copper Industry in the Chilean Economy: An Econometric Analysis
Manuel Lasaga
Copper and Zambia
Chukwuma F. Obidegwu and Mudziviri Nziramasanga
Commodity Exports and Economic Development
F. Gerard Adams and Jere Behrman

Copper and Zambia

An Econometric Analysis

Chukwuma F. Obidegwu
The World Bank
Mudziviri Nziramasanga
Washington State University

LexingtonBooks
D.C. Heath and Company
Lexington, Massachusetts
Toronto

Library of Congress Cataloging in Publication Data

Obidegwu, Chukwuma F.
Copper and Zambia.

Includes bibliographical references.
1. Copper industry and trade—Zambia—Mathematical models.
2. Zambia—Economic conditions—Mathematical models. I. Nziramasanga, Mudziviri. II. Title.
HD9539.C7Z368 338.2′743′096894 81-3761
ISBN 0-669-04659-0 AACR2

Published simultaneously in Canada

Printed in the United States of America

International Standard Book Number: 0-669-04659-0

Library of Congress Catalog Card Number: 81-3761

Contents

List of Figures

List of Tables

Foreword

This book is part of a broader project, sponsored by the Agency for International Development (AID) under contract to Wharton Econometric Forecasting Associates, Inc., on international primary commodity markets and economic development—an integrated econometric analysis of basic policy issues. It presents the study of copper in the Zambian economy and parallels other work on coffee in Central America, the Ivory Coast, and Brazil, and on copper in Chile. The present book builds extensively on the work of various participants in the broader project, and particularly on the work of Derek Ford, Manuel Lasaga, Jirapol Pobukadee, Theophilos Priovolos, and Gabriel Siri.

This project is a continuation of a long line of research at the Department of Economics of the University of Pennsylvania on commodity models and economic development. This research has benefited from support from the Ford Foundation, the Rockefeller Foundation (which financed further development of the commodity models used in this project), the United Nations Conference on Trade and Development (UNCTAD), the World Bank, and other organizations.

Discussions with past and present members of the AID staff and with many other individuals have proved particularly useful in developing the objectives and the structures of this project and in reviewing it as it went along. We note particularly the support and critiques of Keith Jay, Danny Lepzinger, Constantine Michalopoulos, Thomas Morrison, Lorenzo Perez, and William Cline.

We acknowledge with thanks the financial support of AID in connection with this research.

F. Gerard Adams
Jere R. Behrman
University of Pennsylvania

Introduction

The commodity problem has been widely discussed, most notably through the forum provided by the United Nations Conference on Trade and Development (UNCTAD) and the set of agreements between the European Economic Community (EEC) and the member states of the African, Caribbean, and Pacific (ACP) group, as embodied in the first and second Lomé Conventions. The effects of unstable export earnings are assumed to be unfavorable to producer countries, allegedly because the benefits of a price increase seldom seem to offset the disadvantages of a decline in prices on the same scale.[1] First, fluctuations in export receipts negatively affect investment, since the rigidity in the economic structures of most developing countries prevents them from fully utilizing the additional export receipts generated by an unexpected upward shift in export prices. At the same time, an unexpected downward change results in a disruption of investment projects. Second, an unexpected increase in government revenues from exports simply results in an increase in public-consumption expenditures, but there is no equivalent cutback in a downturn. Thus, because of the asymmetric response, a decline in government revenues from exports may be accompanied by increased taxation, government borrowing, and higher inflation rates. Finally, export-earnings fluctuations negatively affect the balance of payments, and the impact of a price decline is made worse by capital flight demands for advance payment for imports, delays in the receipt of export earnings, and other related problems.[2]

These problems are considered prevalent among all commodities, as evidenced by the negotiation in the United Nations of an integrated commodity program, and the creation of a common fund for the stabilization of export receipts of member states.[3] The Lomé Convention, although limiting itself to the stabilization of agricultural raw commodities,[4] nevertheless considered stable metal export earnings to be conducive to expansion of production and introduced an export-earnings insurance scheme against an unexpected significant collapse in output or prices to a single producer. The convention also noted that investment in mineral exploration in Africa had virtually ceased since 1970 because of low prices and smaller linkage effects, and that additional financial and technical resource flows had to be channeled into mining from the EEC countries whose economies depended on external supplies.[5]

Copper mining is the dominant modern-sector economic activity in Zambia, contributing significant proportions of real gross domestic product and government revenues. The importance of copper to the economy is magnified by the fact that it normally provides over 90 percent of the export earnings of Zambia. Copper produced by Zambia is sold in world markets at prices based on the London Metal Exchange (LME) copper-price quotations. The LME price of copper is notorious for its short-term volatility; consequently, Zambian export earnings and government tax revenues from copper fluctuate considerably. This variability of revenues is believed to cause difficulties in the process of economic planning by the government and industry.

The objective of this book is to trace, as quantitatively as the available data permit, the impact of copper mining and the world market price of copper on Zambian economic performance. It focuses on the impact of changes in the copper industry and market on gross micro and macro domestic product, employment, real and nominal stability, international financial position, and sectoral distribution of output. The book presents analyses and tests for the effects of copper-price fluctuations on the performance of the economy and explores the effectiveness of fiscal and monetary policies and commercial policies in the copper industry in counteracting any negative effects of copper-price fluctuations.

This is carried out by dynamic simulation experiments using a macroeconometric model of the Zambian economy. The model features a detailed copper-sector submodel (micromodel) embedded in a macroeconometric model of the economy (macromodel), with explicit two-way linkages between the models. The micromodel explains production, investment, employment and wages of labor, and the demand for intermediate inputs by that sector. These variables are affected by conditions in the world copper market, as summarized by the world price of copper. The macromodel is a multisector model that explains supply and demand for goods and services, employment and wages, domestic price levels, and foreign transactions. Issues that are important but are not modeled quantitatively in the econometric framework are discussed at relevant points in the work.

Historical dynamic simulations with the model are used to investigate the short-term impact of the copper industry and market prices on the economy. These tests involve exogenous changes in copper prices and output and the determination of their multiplier effects. Simulations covering historical and forecast periods are used to test the effects of short-term fluctuations of copper prices on the economy. Policy issues are investigated with historical and forecast-period simulations.

Chapter 1 is a historical analysis of the structure and performance of the Zambian economy and copper industry, focusing on the conditions that led to the growth of the industry and on how the development of the industry has affected Zambia and its people. The discussion includes such issues as government policy toward the industry and the extent of linkages between the industry and the rest of the Zambian economy.

Chapter 2 deals with the issues of export instability and its macroeconomic effects, examining the theoretical and empirical studies on the subject and their application to the Zambian context. Because the empirical studies on export instability have been reviewed in detail elsewhere, our review of it is necessarily short, but the relevant references are provided. The channels through which export instability affects the Zambian economy are discussed. Some of the problems of modeling the effects of export instability and comparing behavior under stable and unstable export regimes are also discussed. Chapter 2 also includes a general overview of the model of the economy used in the study, a discussion of data sources and possible problems connected with the data set and the econometric techniques to be used, and the practical and conceptual problems associated with modeling a developing and dualistic economy. Previous efforts at modeling the Zambian economy are discussed. The chapter ends with a presentation of the general model characteristics, which outlines the general economic framework of the model of the economy.

Chapters 3 and 4 deal with the copper industry (micromodel) and the macromodels, respectively. The structure, specification, and estimated relationships of the copper-industry model are presented in chapter 3, and chapter 4 presents the same for the macromodel.

Chapter 5 deals with model validation, multiplier tests, and tests of the impact of copper-price fluctuations. The model is validated by comparing its predictions to the actual values of the endogenous variables. Multiplier tests are performed in the historical period to determine the sensitivity of the economy, as represented by the model, to exogenous shocks. The chapter closes with a discussion of the tests performed to determine the impact of fluctuations of copper prices on the economy. Basically, these tests compare a model solution obtained using a smooth-trend path of copper prices with solutions obtained assuming various fluctuating patterns of copper prices.

Chapter 6 presents an analysis of some of the critical policy issues in Zambia, elaborating on the social, economic, and political dimensions. Chapter 7 deals with various policies that can be used to counteract the effects of copper-price fluctuations or to promote economic growth, employment, equitable income distribution, and price and real stability. The policies studied include fiscal and monetary policies, commercial policies in the copper industry, and agricultural pricing policy. The efficacy of these policies was studied in each case by comparing simulations of the model embodying the policy with simulations that embody usual or passive policies. The chapter ends with an examination of agricultural pricing policies, which, although unrelated to copper, affect one of the major economic sectors whose performance provides a key to goal attainment in Zambia. Chapter 8 summarizes the results of the study and draws conclusions on policy options.

Notes

1. See "Stabilization of Export Receipts," *The Courier,* No. 31, March 1975, p. 24.

2. Ibid.

3. UNCTAD, *Draft Agreement Establishing the Common Fund for Commodities,* TD/IPC/CONF/L.15, June 1980.

4. The Commodity export stabilization system under the Lomé Convention is known as STABEX.

5. See *The Courier,* No. 58, November 1979, p. 29. The mineral-export guarantee system is to be known as MINEX.

The Zambian Economy and the Copper Industry

Introduction

Zambia is a landlocked nation with an area of some 285,000 square miles and an estimated population of 5.5 million people in 1978. It is bordered by Angola to the West, Zaire to the north and northeast, Malawi to the east, Mozambique to the southeast, and Zimbabwe (formerly Rhodesia) and the Caprivi Strip of Namibia (South West Africa) to the south. The Zambesi River forms a natural boundary with Zimbabwe and the Caprivi Strip.

Zambia, which was formerly the British colony of Northern Rhodesia, became independent in 1964. From the initial major Western invasion in 1889 until 1924, the territory was ruled, in the name of the Crown, by the British South Africa Company under a charter granted it by the British government. The charter also permitted the company to develop the mineral resources of the country under its sole management.[1] The company introduced a system of special licenses, which granted the prospector exclusive rights to specified mineral claims but retained a 50 percent interest in all minerals discovered. Until 1919, most of the licenses were granted to small prospecting companies, which quickly ran out of capital and sold out to larger ventures. The industry remained moribund until the early 1920s, largely because the more accessible copper oxide ores were depleted and the existing technology did not allow for the exploitation of the deeper sulphides. Importation of technology from the United States for the prospecting of deep ores, together with foreign large-scale financing, revived the industry. By 1930, the two major mining companies, the Rhodesian Selection Trust (RST) and the Anglo-American Corporation, owned most of the concessions. The former was controlled and financed by American Metal Climax (later AMAX) of the United States and the latter by the Anglo-American Corporation of South Africa. The Great Depression stopped any further development, and the major mines were closed down until 1934.

In 1924, the British government assumed direct rule of the colony. It appointed a governor, and the settler community elected a legislative assembly. The governor had full authority, however, with the legislature acting as an advisory body. In 1953, the colony joined Nyasaland (now Malawi) and Southern Rhodesia (now Zimbabwe), forming the Federation of Rhodesia and Nyasaland. Nyasaland, like Northern Rhodesia, was administered by the British Colonial Office through a governor, while Southern Rhodesia had been granted internal self-rule in 1929. This federation was to last until 1963.

By 1934, copper production came from the Mufulira and Roan Antelope mines, both underground operations owned by RST, and the Nchanga (open pit) and Nkana mines, both controlled by the Anglo-American Corporation. Other mines, which came into production during the period 1954–1965, were Chibuluma (1956) and Chambishi (1965), both RST operations, and the Anglo-American group's Bancroft mine (1957). As a result of these new operations and expansion at existing mines, total production increased from 138,000 metric tons in 1934 to 240,000 tons by 1943 and 568,000 tons in 1961. The British South African (BSA) Company received royalty payments but paid only 20 percent of the net revenues as taxes to the Northern Rhodesian government. After 1953, these payments went to the federal government, which then distributed the revenues among the three member countries. During the federation (1953–1963), the copper-mining industry of Northern Rhodesia generated the equivalent of £239.2 million in tax revenues. Of this, 48.5 percent went to the federal government, 37 percent to the Northern Rhodesia government, 10 percent to Southern Rhodesia, and less than 5 percent to Nyasaland, as shown in table 1–1. The bulk of the federal government revenues were eventually spent in Southern Rhodesia since it was the more industrialized member of the customs union.

The African population in all the countries had resisted the formation of the federation. This resistance was stronger in Northern Rhodesia and Nyasaland because they resented political domination by the overtly racist Southern Rhodesian settler government. The British government finally gave in to the pressure and appointed commissions to study the issue in the two countries. Both the Devlin Commission (1959) in Nyasaland and the Monckton Commission (1960) in Northern Rhodesia found widespread dissatisfaction. A new constitution for Northern Rhodesia, drawn up in 1961, allowed for the granting of political independence under an African majority government. The federation was dissolved by the British government on 1 January 1964. The National Independence Party, led by Kenneth Kaunda, won the elections in September of that year, and Northern Rhodesia became Zambia in October 1964. Nyasaland had become Malawi.

Independence for Zambia also ushered in a new and unstable relationship between the mining industry and the government of the new country as each tried to maximize its own objectives. Copper prices had begun their upward climb in 1964, and the industry was faced with the possibility of substitution of aluminum for copper were the trend to continue. Zambian producers, along with other major exporters, started using their own producers' price, which was pegged below the London Metal Exchange quotations. The Zambian producers also saw the political change as an opportunity to eliminate copper royalties, considered an inefficient form of taxation, especially since they were based on the now disregarded London Metal Exchange price.

The mining industry had been strongly in favor of the federation, and uncertainty about its future had slowed the rate of investment during 1959–1964.

Table 1-1
Distribution of Taxes from the Northern Rhodesian Mining Industry, 1954-1963
(thousands of British pounds)

	1954	*1955*	*1956*	*1957*	*1958*	*1959*	*1960*	*1961*	*1962*	*1963*	*Total*
From Mining Companies											
Income tax	17,422	16,491	21,421	24,169	12,556	6,643	14,272	17,872	16,315	14,932	162,113
Customs duties	220	220	275	242	138	114	165	159	173	150	1,856
From BSA Company											
Mineral tax	1,762	2,714	2,762	1,795	1,365	2,568	2,786	2,607	2,548	2,684	23,591
Royalty tax	2,869	2,686	3,693	4,446	3,148	2,174	3,399	4,600	3,982	4,253	35,250
Employee income tax	1,024	1,202	1,675	2,016	1,280	1,372	1,778	1,965	2,060	2,096	16,468
Total	23,317	23,313	29,826	32,668	18,487	12,871	22,400	27,203	25,078	24,115	239,278
To Federal government	11,735	11,266	14,778	16,819	9,158	5,418	10,413	13,014	11,968	11,389	115,958
To Northern Rhodesia	8,246	8,874	10,854	11,027	6,604	5,651	8,797	10,161	9,428	9,224	88,866
To Southern Rhodesia	2,283	2,171	2,871	3,304	1,895	1,326	2,238	2,827	2,586	2,460	23,961
To Nyasaland	1,053	1,002	1,323	1,518	830	476	952	1,201	1,096	1,042	10,493
Northern Rhodesian revenue as percentage of total	34	38	36	34	36	44	39	38	38	38	37

Source: *Northern Rhodesia Chamber of Mines Yearbook 1963*, Table 4, p. 23.

After 1963, the mining corporations blamed the lack of new mineral developments on the royalty system. They argued that the incremental output would have to come from higher-cost ore deposits and that royalties did not encourage the necessary investment because they did not take production cost differentials into account. Royalties based on the London Metal Exchange price had cut into profitability, particularly during 1964–1966, when the corporations were selling their output at lower producer prices.[2] Initially, the government had been willing to discuss the issue of royalties and had noted their effect on investment. However, as metal prices continued to increase through 1966, with no appreciable output response, the attitude changed. It was then believed that royalties could not have been the major reason for the lack of investment, since the corporations had previously invested a substantial share of their earnings, even as they paid royalities to the British South Africa Company, without complaint.[3] This disagreement over the determinants of output expansion and the appropriate form of taxation was to lead eventually to the nationalization of the industry.

There were other problems as well, one of the most important being the availability of skilled labor at the current wage rates. Following independence in 1964, the status of the European mineworkers was changed to that of expatriate. They were then classified into two groups: those whose contracts were to have a life expectancy of two years or more and those whose positions were subject to immediate Zambianization. The resulting uncertainty, especially of the latter group, led to higher turnover rates, difficulties in recruitment of skilled expatriate labor, and skilled-labor shortages.[4] Education of the indigenous population was grossly neglected prior to independence. It is estimated that at independence the country had fewer than 100 university graduates and 1,500 high school graduates. The country therefore depended on expatriate labor in all sectors of the economy. Their repatriated earnings constituted a foreign-exchange cost. Production continued to rise to 685,000 tons in 1965, however, largely because of output from the new mines and an expansion program at Mufulira mine. The mineral rights and royalties had reverted to the Zambian government at independence, and the increased production, coupled with higher copper prices, resulted in record export revenues accruing to the new government. The industry was now taxed as follows: (1) a royalty equal to 13.5 percent of the London Metal Exchange copper price, less 16 kwacha per ton of copper produced; and (2) an income tax of 40 percent of net profits, changed to 47.5 percent in 1965 and 45 percent in 1966. This rate applied to corporations with net incomes of over 1 million kwacha. The rate for smaller mines was 35 percent. The mining companies protested the royalty as "an open-ended financial risk" and pressed for a revision of the tax laws. They became reluctant to withhold earnings for investment and paid out an average of 94 percent of their earnings as dividends, as opposed to 50 percent in 1961. The tax laws were eventually revised in 1966. The government of Zambia imposed an export tax at the rate of 40 percent of the amount by which the monthly average London Metal Exchange price exceeded 600 kwacha per long ton. This raised additional revenues in 1966, but its effect was countered by a drop in sales in 1967.

In 1968, the government restricted the repatriation of earnings to 50 percent of the declared dividend or 30 percent of their paid equity, whichever was less, but the retained revenues simply accumulated as reserves in the Central Bank. In 1969, the royalty and export-tax laws were changed. In their place was imposed a mineral tax of 51 percent of profits, with the balance subject to the usual income tax. The effective tax rate was thus 73.05 percent of profits. At the same time, the copper-mining industry was partially nationalized, with the government purchasing a controlling 51 percent of the outstanding shares. The industry was consolidated into two groups. The former Anglo-American Corporation holdings formed Nchanga Consolidated Copper Mines (NCCM), while RST properties became Roan Consolidated Copper Mines (RCM). The state formed a holding company, the Zambia Industrial and Mining Corporation Limited (ZIMCO), which then held the government's 51 percent share in both NCCM and RCM. To finance the purchase, ZIMCO issued bonds at 6 percent, payable in 1978 for RCM and in 1982 for NCCM, for a total acquisition cost of U.S. $331 million. The former owners maintained control over operations, and the maximization of profits was retained as the overall goal of the industry.[5] Anglo-American Corporation and Roan Selection Trust also obtained a substantial management contract to operate the industry. Their management remuneration was based on the gross value of output, profitability, and the number of skilled expatriate workers recruited from abroad. All restrictions on dividend repatriation were removed.

Government dissatisfaction with the rate of expansion of output and the level of localization of the labor force continued, however. In addition, the relatively high metal prices in 1973 generated substantial revenues for the minority shareholders from the management contract, and these were repatriated. In August 1973, the government of Zambia redeemed all the outstanding ZIMCO bonds. In 1974 and 1975, the management contracts with Anglo-American and RST were terminated for a lump sum of U.S.$78.6 million, and both RCM and NCCM became self-managing. The negotiations and payments were not completed until 1977.

Zambia's landlocked location has presented special transportation problems. Prior to 1966, most of the copper exports to Europe and Japan and the imported inputs were carried by rail through Rhodesia to the Mozambique port of Beira. A smaller proportion of copper exports were shipped on the Benguella railway through Zaire to the port of Lobito in Angola. In 1965, Rhodesia unilaterally declared independence from the United Kingdom, an act considered illegal by Zambia. Opposition to the regime and the imposition of sanctions on Rhodesia by the United Nations led to a closing of the border between the two countries. This caused a transportation bottleneck. At first, copper exports were airlifted out of Zambia, and planes brought in fuel and other imported inputs on the return trip. Then, after 1966, alternative routes were opened up, by road to Dar es Salaam, and Mombasa as well as Malawi. Imports came by rail through South Africa and Botswana. Zambia developed her own coalfields to replace Rhodesian sources.

The result was a 30 percent increase in copper-transportation costs between

1964 and 1968. Thereafter, they remained almost constant. Production seems to have been significantly affected only in 1966. Thereafter, transportation problems seem to have caused only temporary disruptions in the movement of exports. Completion of the railway line between the copperbelt and Dar es Salaam partially relieved the problem, but it has been plagued, in turn, by equipment problems and capacity constraints. The effect of this uncertainty in transport facilities has fallen most heavily on imported inputs and has resulted in higher costs. With the unilateral declaration of independence of Rhodesia in 1967, the industry substantially increased its desired level of coal stockpiles. The value of stocks of other inputs held by the RCM division rose by 37 percent in the same year, while the pipeline stocks of copper exports increased by 20,000 tons above the normal level.[6] Since these copper stocks were already the property of the importers, they did not affect total export figures, but they did affect the confidence of the industry in meeting sales deadlines. The transportation constraint also caused a substantial variability in the port charges at both Beira and Dar es Salaam.

Zambia is one of the largest producers of copper in the world, and copper exports are the mainstay of the economy. Zambia produces other minerals, principally zinc, lead, and cobalt. These are rather insignificant, however, compared to copper, which comprises about 96 percent of mineral output by value. In addition to mining, agriculture is a significant economic activity. It produces food for domestic consumption and cash crops for domestic secondary industries and exports. Recently, agricultural production has not met domestic demand, food imports have increased, and exports of agricultural commodities have declined. There is a growing manufacturing sector, generally engaged in the production of consumer goods.

The Zambian government plays an active role in the economy, virtually controlling the pricing and distribution of traded agricultural products. Public ownership of industrial firms is substantial; many firms are wholly owned by the government or are owned jointly with private interests. In 1969 and 1970, the government acquired controlling interest in the major privately owned and foreign-owned firms. Foreign investment is welcome, but joint ownership between foreign investors and the government is encouraged.

Following independence in 1964, Zambia experienced rapid economic growth stimulated by government expenditure on services and capital formation. This spending was financed by revenues generated by high copper prices. Foreign-exchange revenues provided for the importation of the required capital and skilled labor. Particular emphasis was placed on such areas as education and health, infrastructure facilities, and the development of manufacturing industries. After 1970, copper prices declined, and, with little or no growth in output levels, there has been a decline in the growth rates of the manufacturing and service sectors, the two sectors most affected by government spending. With the exception of the commodity price boom in 1973–1974, foreign exchange has become a binding constraint on Zambian economic development. The external debt has

Table 1-2
Growth Rates of Selected Variables

	Average Annual Percentage Growth Rates		
	1966-1976	*1966-1970*	*1971-1976*
GDP	8.7	11.4	6.8
GDP, real	2.4	2.0	2.8
Contribution to real GDP			
Mining and quarrying	−2.0	−4.7	0.3
Agriculture	2.7	1.7	3.5
Manufacturing	7.7	11.2	4.8
Services	6.6	9.3	4.3
Exports of goods and services	10.5	15.8	6.1
Imports of goods and services	10.7	13.3	8.5
External debt outstanding	23.7	34.6	15.4
Debt-service payments	48.0	44.0	50.7
Total population	3.0	2.9	3.0
Urban population	7.6	8.5	6.9

Sources: Republic of Zambia, *Monthly Digest of Statistics,* Lusaka, several issues and World Bank, *World Tables, 1980.*

risen considerably. In 1976, the debt service was 9 percent of exports, compared to 2 percent in 1966; by 1978, debt service had risen to 23.5 percent. Table 1–2 summarizes the growth rates of the gross domestic product and other selected variables in the Zambian economy during the period 1966–1976, and Figure 1–1 shows the distribution of real gross domestic product by primary, secondary, and tertiary sectors.

The Performance of the Zambian Copper Industry

Until 1970, Zambia was the third largest producer of copper in the world, following the United States and the USSR. Recent statistics indicate that this position has been usurped by Chile and Canada. Zambia is the second largest exporter of copper in the world (after Chile) and the largest exporter of refined copper. In addition to mining, smelting, and refining operations, a small copper-fabricating plant, ZAMEFA, began operation in 1974, with output intended for domestic consumption.

The production of copper rose from 6,300 metric tons in 1930 to 709,000 metric tons in 1976. Figure 1–2 shows Zambian copper production from 1953 to 1976. The compound percentage growth rate of output was 7.7 percent from 1950 to 1959, 2.5 percent from 1960 to 1969, and 0.6 percent from 1970 to 1976. The growth rate of output from independence to 1976 was only 1 percent. These figures indicate that the growth rate of copper production has declined rather dramatically since independence, especially after 1970. Since then the industry has been unable to meet its ambitious production targets.

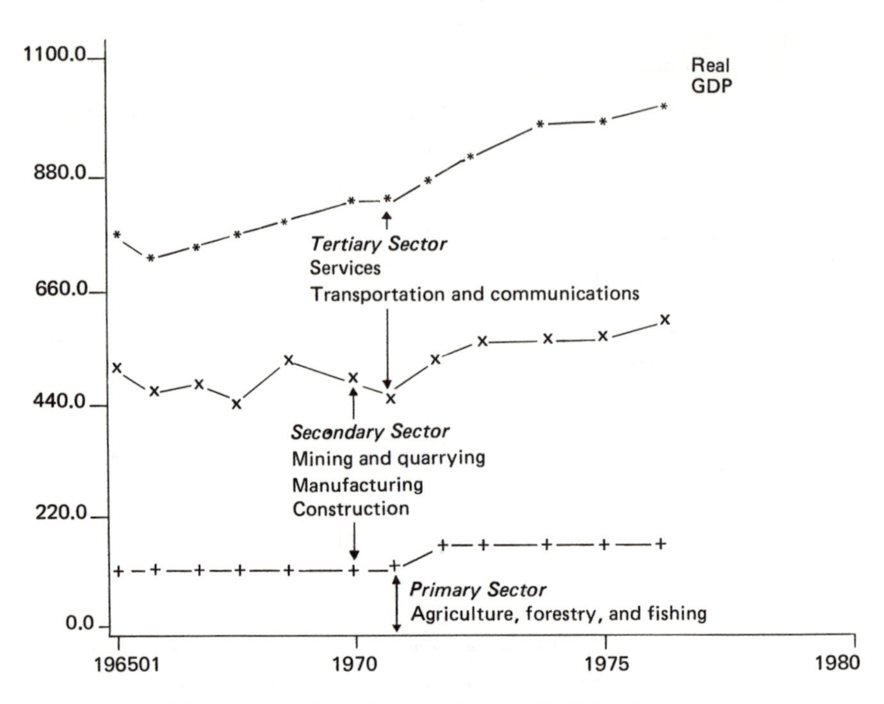

Figure 1-1. Distribution of Real GDP by Sector

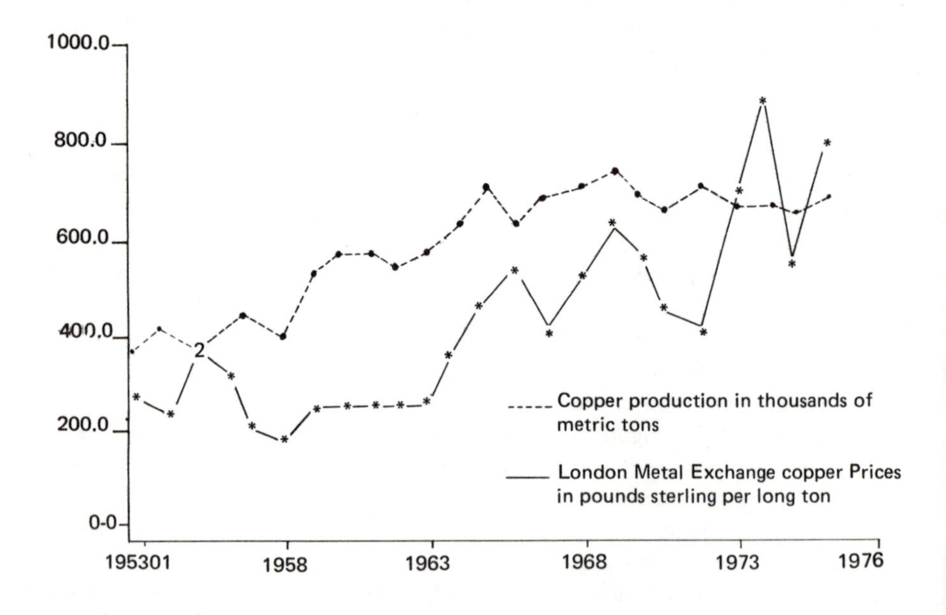

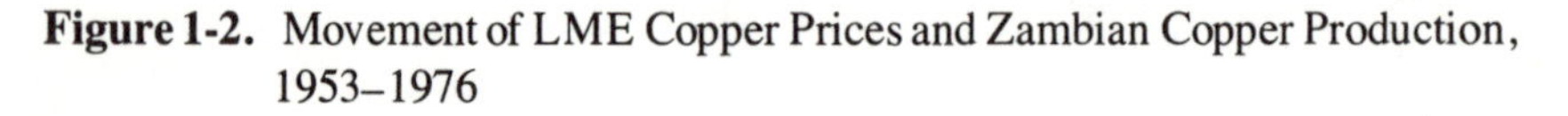
Figure 1-2. Movement of LME Copper Prices and Zambian Copper Production, 1953–1976

In 1969, just before the takeover of the industry by the government, production for 1975 was projected at 793,000 metric tons. Expansion plans were later revised upward, and the new production forecast for 1975 was 900,000 metric tons.[7] This represented an average annual growth rate of production of over 4 percent from 1969. The realized output for 1975 was 677,000 metric tons, far below the target figure. The projections of production increases may have been too ambitious. Although a higher growth rate of output had been achieved in the past, the conditions under which it was achieved were vastly different from the circumstances surrounding the industry during the plan period.

From 1959 on, as a result of uncertainty in the political situation, the mining companies invested cautiously and paid out most of their profits as dividends to their foreign owners. The Zambian public regarded the companies as insensitive to the needs of the country. It appeared that the mining companies accepted the inevitability of a government takeover, which had an influence on their investment programs. The takeover did resolve some of the uncertainty about their future, but it introduced new uncertain situations. The new partnership introduced more complicated agency relationships, which could not survive in the long run. It seemed clear that the new partnership was a transitional arrangement to help Zambia avoid a costly disruption of its mining operations, and there was no reason for the mining companies to change their commercial objectives. Their primary interest remained the maximization of short-run profits and high-dividend payments. It therefore seems reasonable that their plan before the takeover represented their best assessment, given expected market conditions, of the level of output that would be feasible and profit-maximizing. The companies also realized that high-level investments would decrease dividends, since the majority owner, the government, would favor financing investments out of profits.

It would appear, then, that there might in fact have been two plans in operation—one explicit, complying with the government goals, and the other implicit, corresponding to the goals of the mining companies. Subsequent events tend to support this view. Soon after the takeover, the government's dissatisfaction with the operation of the partnership became evident. The mining companies were accused of being reluctant to plow profits back into expansion, much as had been the case before the takeover. This was presumably the major reason the government decided, in 1973, to assume effective control of the industry. The actual assumption of effective control by the government was not accomplished until early 1975. The explicit plan was eventually embodied in the Second National Development plan (SNDP), which covered the period 1972–1976. At the end of this plan period, the intended investment expenditures of the copper industry were underspent by about 20 percent.[8] However, all this underexpenditure cannot be attributed solely to the existence of an implicit plan, as we shall discuss.

In the two years following the takeover, copper prices declined dramatically. This limited the availability of capital for expansion projects. Profits declined further, limiting the funds that might have been generated from internal resources. External financing also became more difficult to raise. The Zambian government had to make significant compensation payments for its share of the industry and thus could not divert its dividends to expansion. Copper prices improved in 1973, but the government borrowed from foreign sources in order to retire the compensation bonds, thereby reducing its capacity to borrow more from abroad when copper prices declined in 1975. Real investment expenditures in mining declined from 1974 to 1976. The availability of finances may have been a factor in this decline.

Political events in Southern Africa have had adverse effects on Zambia and its copper industry. The unilateral declaration of independence (UDI) by Rhodesia, the political unrest in Zaire and Angola, and the racial policies of the South African government all created problems for Zambia. The most severe appears to have been UDI, given the historical links between Zambia and Rhodesia. The most direct impact of UDI on the copper industry was on transportation and intermediate input costs. The mining industry had to substitute locally mined coal for the more efficient Rhodesian coal when the Rhodesian authorities drastically increased the price of the coal sold to Zambia. This move toward a reliance on local coal was inevitable, however, in view of the uncertainty of the availability of Rhodesian coal following UDI. Zambia continued to use the transportation route through Rhodesia to Beira, although at much higher fees, until 1973, when the border between the two countries was closed. An alternative route through Zaire to Lobito in Angola was closed in 1975 because of internal problems in Angola. However, because of political uncertainties, neither of these routes was reliable even when they were open. With the closure of the Lobito route, the port of Dar es Salaam became the only one available for Zambian foreign trade.[9] Until a railway was completed, the link between Zambia and Dar es Salaam was a substandard roadway, which was inadequate to handle Zambian traffic.

Hence, Zambia was forced to continually reroute her imports and exports. The greater impact of this exercise was on imports, which were larger than exports by volume. However, exports were subject to delay when transportation was available. The lack of sufficient capacity for the transport of imports resulted in shortages of materials and equipment needed for new and ongoing mining projects. These transportation uncertainties forced mining companies to carry higher levels of inventories; this together with higher transportation costs increased operating costs. New and more expensive sources of material supply had to be found because some new transportation arrangements precluded imports from lower-cost South African markets.[10] In addition, continued trade with South Africa was indeed politically embarrassing to Zambia in view of her stand against apartheid.

The political situation in Southern Africa had other effects on Zambia. It may have scared away potential foreign investors who would have wished to explore for minerals. Zambian industry still relies heavily on expatriates for skilled labor. The ability of the industry to attract qualified personnel from overseas may have been affected by the perceived increased risk of living in Zambia caused by the political situation. In addition, Zambian expatriate wages may not have been competitive in the world market for experienced mining personnel, especially in view of the perceived high risk of living in Zambia. As a result, in recent years, the industry has experienced a high turnover of expatriate employees. The experience level of expatriate employees has declined, but there is no evidence that this was a deliberate policy. There have been shortages of key people, especially experienced maintenance engineers. These shortages probably explain the increased frequency of the occurrence of technical problems and equipment breakdown in the industry, causing delays in production.

Other events in the mining industry have hindered production growth. The most serious of these was the cave-in of an underground mine at Mufulira in late 1970. Before the accident, this mine carried the lowest cost in Zambia and was the largest underground mine. The effect of the accident was a loss in production estimated to be about 125,000 metric tons in the first year. Rehabilitation of the mine has taken longer than expected, and by 1976 output was still at about two-thirds the preaccident level. Plans to increase output at the Konkola mine were not realized because of persistent dewatering problems. The ore grade of Zambia mines has been declining so that more ore has to be mined for each ton of finished output. This results in lower production from available capacity and increased costs per unit of output. Domestic and worldwide inflation has contributed additional cost pressures to the Zambian industry, further reducing profitability.

The performance of the mining industry must be disappointing to the Zambian government. The government had attempted to use its influence to facilitate the expansion of output, first by what amounted to moral persuasion and political pressure, and finally by direct involvement in the industry. Neither of these efforts seemed to have been very successful. Admittedly, there were several other factors working against Zambia, but the takeover may have had a detrimental effect on the performance of the industry. Although this cannot be judged by short-run results, after eight years, no case can be made for a trend toward improved performance. There seems to be a need for decisive action by government and a clear statement in such areas as industrial and commercial policy, organization redesign, and management incentives. Running the old company under a new name, with even less clear objectives and policies, is unlikely to meet the high goals the government has set for the industry.

The Pricing of Copper

Zambia sells its copper at prices based on the London Metal Exchange (LME) prices. Zambian producers experimented with an administered producer price from 1964 to 1966. This price was set by Zambian producers, in conjunction with other major copper exporters, to stabilize the market and to discourage the growth of substitute materials. In 1966, the LME prices increased sharply, widening the gap between the producer price and the market price. The producers were forced to abandon their price in favor of the LME price because the loss of revenues was too great to be borne and the price stabilization effort did not appear to be working. In addition, the royalties producers paid to the government were based on the LME price.

The LME price of copper is subject to extreme, short-run fluctuations. These fluctuations are a function of changes in stocks, speculative activity, and shifts in demand and supply. The long-run equilibrium price of copper is influenced by the trends in demand and supply of copper and its substitutes. Up to 1976, the highest average annual price ever attained was £878 per long ton in 1974, which corresponded to the commodity price boom in 1973–1974. However, during that year the price varied between a monthly average high of £1,297 in April to £552 in December, a 60 percent decrease in eight months. Since 1970, the lowest average annual price was £428 per long ton in 1972. Figure 1-2 shows average annual values of the LME copper prices from 1953 to 1976.

The extreme volatility of copper prices translates to wide, short-run fluctuations in Zambian export and government revenues. These fluctuations are presumed to create difficulties for planners in government and industry, even abstracting from any other adverse effects they may have on the economy. In addition to supporting multilateral efforts to stabilize commodity prices through the United Nations, Zambia is a member of the Intergovernmental Council of Copper Exporting Countries (CIPEC). The objective of this body, whose membership consists of developing nations that export copper, is to influence price by controlling supply. An important part of this objective is to prevent the excessive fluctuations in copper prices.

However, CIPEC has not been successful either in setting prices to reap monopoly gains or in establishing prices around their long-run market-equilibrium levels. There have been many explanations for this inability to intervene successfully in the copper market.[11] CIPEC members account for only 30 percent of total world output, although they contribute over 60 percent of world exports. Any reduction in shipments by CIPEC members would be met either by increases in primary production in nonmember producer countries or by increased use of scrap and substitute products. These factors would imply a very elastic long-run demand curve for CIPEC copper, so that increases in price would lead to a decline in export revenues. There are organizational problems involved with running CIPEC as a successful cartel. The interests of individual members may be better served by increasing production when the price falls so as to keep up export revenues, which would be an incentive to cheat. Pindyck used a model of the copper market to study potential gains from cartelization.[12] He concluded that gains were negligible for copper, so that, even if CIPEC were able to solve its organizational problems, it would still not be able to achieve monopoly gains.

The Role of the Copper Industry in the Economy

The mining industry is the dominant sector in the Zambian economy, and copper is its most important mineral. Value added by this sector forms a significant portion of the gross domestic product (GDP), although this portion has been declining in recent years. The contribution of the mining sector to GDP is shown in table 1–3 and figure 1–3.

Table 1-3
Contribution of the Mining Sector to Gross Domestic Product

	Contribution to GDP at Market Prices		*Contribution to Real GDP*	
	Millions of Kwacha	*Percentage of GDP*	*Millions of 1965 Kwacha*	*Percentage of Real GDP*
1965	290.2	38	290.2	38
1966	379.2	41	242.8	33
1968	410.8	36	221.2	28
1970	457.1	36	230.2	27
1972	317.7	24	214.5	24
1974	626.0	33	227.3	23
1975	145.0	9	207.0	21
1976	195.0	11	228.0	23

Source: Republic of Zambia, *Monthly Digest of Statistics,* Lusaka, several issues.

Figure 1-3. Mining Value Added as a Percentage of Real and Nominal GDP

The decline in the contribution of the mining sector to the GDP is because of the sluggish performance of the industry since 1970 (just discussed) and the deliberate policy of diversification of the economy. The latter has resulted in high growth in other sectors, notably, manufacturing and services. This effort uses resources generated by copper through taxes and export revenues. The rapid decline of the proportion of mining output in nominal GDP in 1975 and 1976 was primarily caused by the low international prices of copper.

The mining industry is relatively capital-intensive. Between 1970 and 1975, wages and salaries made up 38 percent of the total costs of production of refined copper and only 14 percent of its total value. This compares with 30.9 percent of total value for the rest of the economy, obtained by Fry and Harvey.[13] Over the years, there has not been much change in total employment in copper mining, although total physical output has risen. In 1960, total employment and output was 47,300 people and 576,000 metric tons, respectively; the corresponding figures for 1970 were 48,500 and 684,000. However, since 1970, employment has been rising steadily, reaching 59,300 in 1977, while output has not increased accordingly.

The increase in labor productivity before 1970, as measured by the physical output-labor ratio may have been caused by the increased use of capital and an improvement in the skills and training of the labor force, particularly of the Zambian employees. The decline in turnover among local employees may also have been an important factor. However, after 1970, although the capital-output ratio continued to rise and local labor-force skills may have improved, probably at a diminishing rate, other problems dominated any possible gains in labor productivity. The quality of supervision and management may have declined, because of the lower level of experience of both expatriate and Zambian managers and supervisors and the increased turnover of expatriate employees. It must also be noted that, since 1970, realized output has always been less than planned output for reasons that were discussed in the previous section. The higher level of employment may actually reflect levels needed to achieve planned output targets. Finally, any conclusions concerning changes in productivity must be qualified by the possible errors of measurement. Our data on employment are in man-years and therefore give no indication of the intensity of labor utilization.

Zambia is dependent on the earnings of its copper exports for foreign exchange. Copper contributes over 90 percent of the total value of exports. The ability of the country to import investment goods to develop and diversify the economy depends largely on its production of copper and the prevailing prices on the international market, and there is little prospect of reducing this dependence in the foreseeable future. Zambia's principal nonmineral exports are tobacco and maize. The tobacco industry has not been performing well and has in the past received government subsidies. The production of maize is subject to weather variations; during some dry years, Zambia has imported maize to make up for shortfalls in its domestic production. Other agricultural commodities—for example, rice, groundnuts, and cotton—could be developed for export, since Zambia possesses both good agricultural land and a suitable climate. The development of these alternative exports, however, will take time and resources. It is a real dilemma: Zambia needs to lessen its dependence on copper but must continue to invest in expansion of copper production so as to obtain resources to carry out the necessary diversification.

The mining industry contributes a significant proportion of government revenues (see table 1–4). This proportion has been declining because of the slow growth of output since 1970, the fact that copper prices have not been as favorable as they were in the 1960s, and rising costs in the industry, which have led to lower profits. An effort on the part of the government to diversify its sources of revenues has also been partly responsible for the relative decline of the contribution from copper. Figure 1–4 shows the movement of mineral and nonmineral revenues between 1964 and 1976, and figure 1–5 shows indexes of total government revenues, mineral revenues, and LME copper prices. These graphs show that there is a tendency for mineral revenues, total revenues, and copper prices to move together; this tendency is much stronger after 1970. Mineral revenues show much higher trend-adjusted variance than do total revenues or copper prices, and the variation in mineral revenues is the greatest source of trend-adjusted variance of total revenues.

Government Policy and the Mining Industry

Since independence, government policies toward the mining industry have been dictated by the need to make it more responsive to Zambian development. Government policies affect the industry in such areas as ownership and control, taxation, investments and financing, and manpower policy.

Table 1-4
Contribution of Mineral Resources to Total Government Revenues

	Government Revenue in Millions of Kwacha	*Contribution to Government Revenue*	
		Millions of Kwacha	*Percentage*
1963[a]	72	25	34
1964	108	57	53
1966	225	163	64
1968	306	183	60
1970	432	218	52
1972	315	69	22
1974	650	341	34
1975	448	59	13
1976	443	12	3

Source: *Zambian Mining Yearbook,* various issues.

[a]During federal years; includes part of federal revenues spent in Zambia.

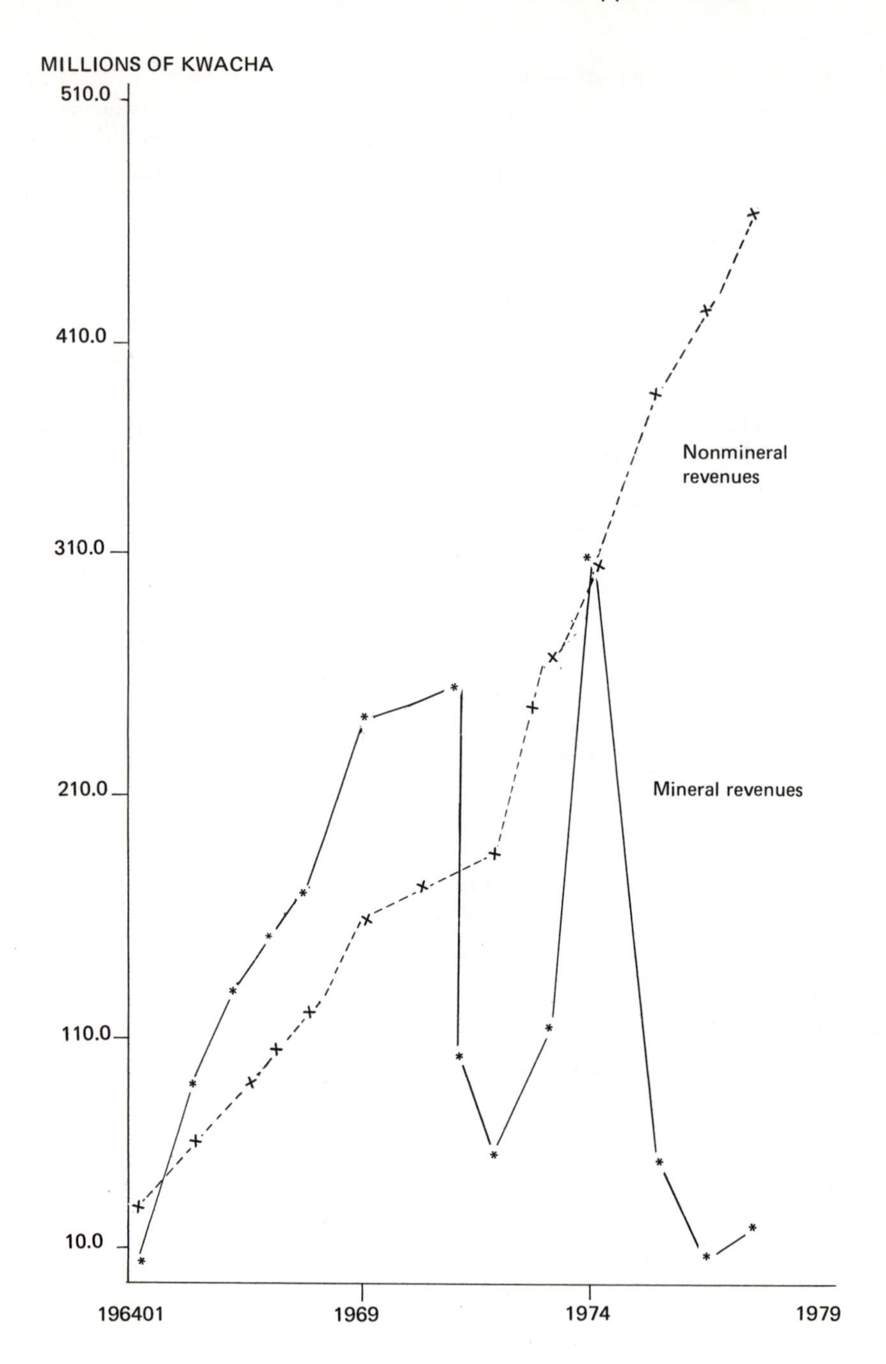

Figure 1-4. Government Mineral and Nonmineral Revenues

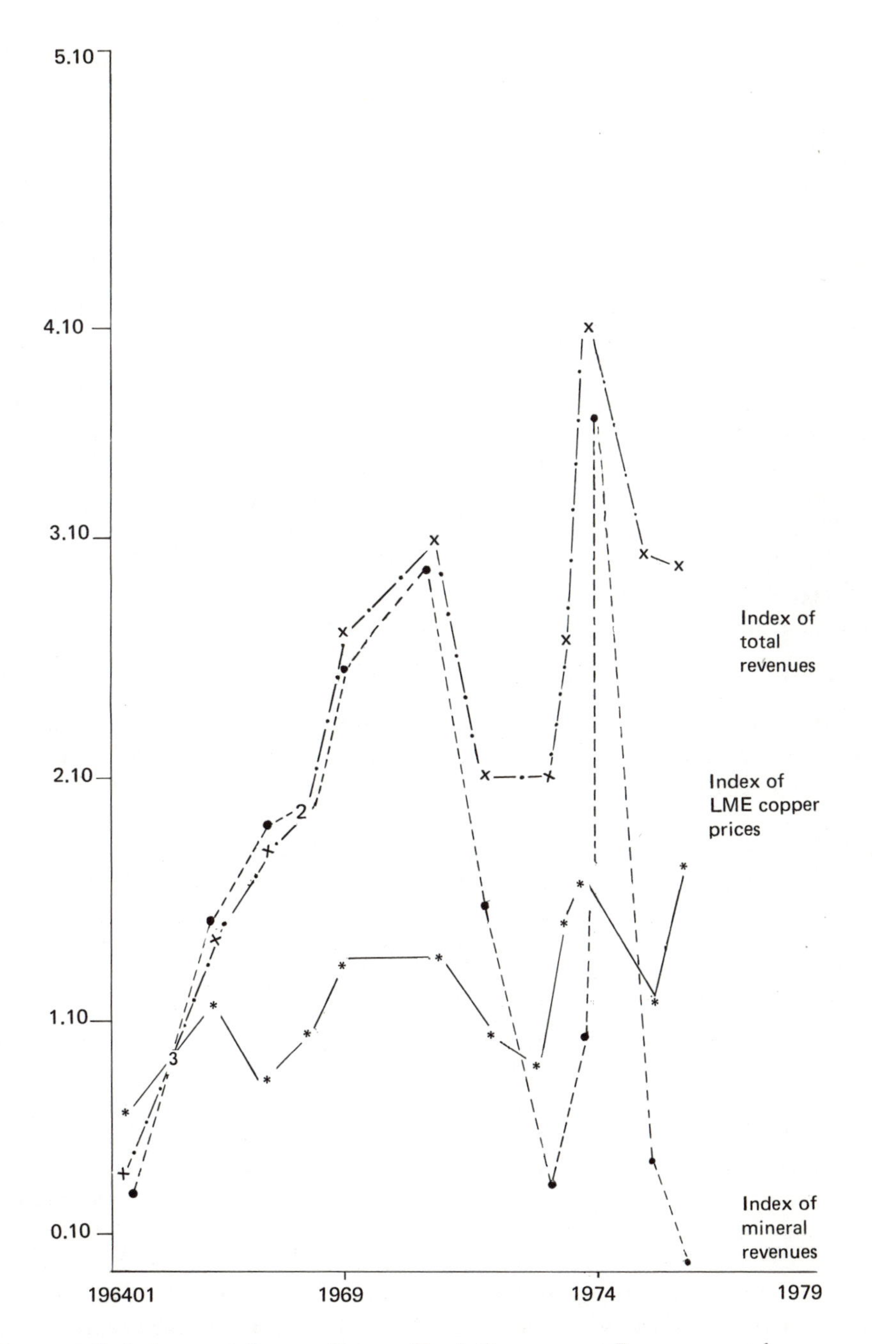

Figure 1-5. Indexes of Copper Prices, Total Government Revenues, and Mineral Revenues (1965 = 1.00)

In 1970, the government became a majority owner of the mining industry, and, by February 1975, it assumed full and effective control of the industry. The takeover was motivated by political and economic considerations. The mining industry had a reputation for being insensitive to the welfare of Zambia. There was also a feeling among Zambians that mining was too important to their future to be left to absentee owners. In Zambia, only the government had the resources with which to buy a majority interest; the possibility of private Zambian citizens participating financially was never really under consideration. A government takeover was consistent with the economic actions the government had taken in the previous year, when it acquired controlling interest in major industrial and commercial enterprises.

Government taxation policies are aimed at maximizing government revenues from mining while providing the industry with incentives to expand production and employment of Zambians. After independence, mining companies had complained that royalties and export taxes discouraged expansion of production in high-cost mines, since they were based on the volume of output and the world market price and not on profits. The tax changes of 1970 contained generous provisions in the treatment of captial expenditures for tax purposes, which was done to encourage expansion. Capital expenditures on new projects could be written off in the year they were made. This law may have accelerated the capital-intensive development of low-grade ores and tailings at Chingola as part of the expansion program. It also may have contributed to the high marginal capital-labor ratio at other mines, thus dampening the employment gains from higher production. In 1973, the capital-consumption allowance was again based on the estimated life of the equipment.

Before takeover, the government could only influence mining investment through economic incentives. Since 1973, the role of the government has become more direct, and investment in the industry is now linked to the economic condition of Zambia as well as to world copper market conditions. The government could now opt for a low-dividend policy. This would increase domestic saving, as lower dividend payments would be sent abroad, and would reduce the need for the industry to borrow on the international financial market. However, this kind of policy is obviated by the need to attract venture capital from abroad for exploration and development of new mines. This need for venture capital implies that government foreign-exchange restrictions should not impede the flow of dividends abroad. As part of the 1970 takeover agreement, the government accepted the responsibility of guaranteeing all loans raised abroad by the industry, thus tying the financing of copper development from external loans to the overall health of the Zambian economy. Since then, foreign loans and lines of credit have become a major source of investment funds, partly because of this shift in ownership and partly because of low copper prices, which reduced profits.[14]

Government policy affects the use of manpower in the industry. Since independence, the government has pursued a policy of Zambianization, which aims to promote Zambians to supervisory and managerial positions in both industry and government. At the time of independence, most of the industry in Zambia was foreign-owned. It was hoped that the presence of Zambians in managerial positions would make industry responsive to the needs of the country, while reducing the foreign-exchange costs of employing expatriates. Although this policy applies to all industry, government, and parastatal firms, its implementation in the mining industry has received the most scrutiny. The mining companies provide training programs for Zambian employees. In addition, they sponsor employees or prospective employees in technical colleges and universities in Zambia and abroad. A consequence of this policy is that expatriates are employed only if no qualified Zambian can be found, and then on renewable two-year contracts. The implementation of Zambianization has been slow; the industry has been accused of not moving fast enough, but the availability of Zambians with the formal education required for most of the technical and managerial positions has been a severe constraint. In addition, the industry has had to compete with the government and other industries for the few available people.

The employment of expatriates is a drain on the foreign-exchange resources, as they must be allowed to remit some of their income abroad, and some of their bonus payments are made in foreign currency. To discourage the unnecessary hiring of expatriates, a selective employment tax of 15 percent is imposed on the wages and salaries paid to non-Zambian personnel. There is also a flat education levy, introduced in 1977, for which every company in Zambia is liable. The proceeds of this levy go toward education and training. Zambianization by itself would not yield a significant number of new jobs, but it would increase the quality of employment. Similarly, the number of new jobs created by expansion will depend on the degree of mechanization of the new operations. A major concern is upgrading the training and skills of the labor force, since the availability of skilled labor is a constraint on the growth of the economy.

These are the major areas where government policy has a direct impact on the mining industry, but there are several others. The government sets the rates paid by the mining companies for the use of Zambian railroads. Its control of import and export routes, as well as its controls on imports and foreign exchange, affects the ability of the mining companies to import intermediate and capital goods and to employ expatriate labor. The government owns the rights to mining in Zambia and grants, extends, and revokes concessions to mining and exploration.

Linkages between the Copper Industry and the Rest of the Economy

Before independence, the copper industry had minimal impact on the local economy, but stronger links have now been developed through government ownership and control. The most direct links are through export revenues, gov-

Table 1-5
Value of Domestic Inputs into the Copper-Mining Industry of Zambia, by Sector of Origin, 1965-1967 and 1969
(thousands of kwacha, current prices)

Sector	*1965*	*1966*	*1967*	*1969*
Agriculture[a]	200	240	240	2,110
Mining	4,340	5,886	13,507	27,700
Wood and wood products	2,360	1,566	513	2,600
Chemicals	820	900	2,434	3,360
Basic metal products	10,580	10,726	12,567	–
Machinery	1,080	1,260	3,800	10,340
Construction	15,160	15,962	16,322	25,150
Electricity and water	14,440	13,934	15,325	16,780
Distribution	8,700	7,324	10,125	32,500
Banks and insurance	2,400	3,760	5,672	20,290
Transportation	2,800	1,954	5,359	4,380
Total domestic inputs	62,880	72,506	98,395	178,220
Total imported inputs	37,120	31,478	37,893	27,940

Source: Republic of Zambia, *National Accounts 1965-66, 1967, 1969* (Lusaka: Central Statistical Office, 1966, 1967).
Note: Statistics for 1968 are not available.
[a]Includes foods and fibers but not timber and timber products.

ernment expenditures, and employment of intermediate inputs. Zambia, like most other developing nations, imports most of the capital goods for the development of all sectors of its economy. In addition to capital goods, Zambia has to import intermediate inputs for its import-substitution industries and consumption goods to meet the demand for foreign goods by the urban sector. As the major export, copper provides most of the necessary foreign exchange. The copper industry has spurred the growth of other industries that provide services and input to it. Table 1–5 shows the value of domestic intermediate inputs, at current market prices, purchased by the copper industry for the periods 1965–1967 and 1969. These are the only years for which data are available. According to the statistics, 62 percent of total intermediate inputs in 1965 came from domestic sources. This figure rose to 70 percent in 1966 and 86 percent by 1969. The data substantially overestimate the actual link between copper mining and the domestic industrial sector. Apart from the impact of differential price increases between domestic and imported inputs, domestic inputs have not been adjusted for their indirect-import content. These are important for the construction, basic-metal, and transportation industries, because Zambia does have a steel industry. The substantial increase in 1969 of inputs from the financial and distribution sectors was largely the result of restrictions placed on the repatriation of dividends. This contribution has since declined.

Despite these qualifications, however, the copper-mining industry has spurred the growth of other industries to provide it with inputs and services. Zambia's coal deposits were developed in 1963 to substitute for imports from

Southern Rhodesia (now Zimbabwe), which led to the sharp increase in 1969 in the value of inputs from the mining industry itself. The backward linkage with the agricultural sector is relatively weak. Wood and wood products used in mining consist mainly of props for underground mines. The chemical industry supplies explosives, but they have a relatively high import content, as do inputs from the basic-metal and machinery sectors. Inputs from construction and electricity probably have the highest domestic content of nontradable inputs, with the construction sector providing housing for mining labor. Electricity comes from the Central African Power Corporations grid, which is owned jointly by Zambia and Zimbabwe. Most of the production is provided by the Kariba hydroelectric scheme, with the balance from the mining industry's own thermal power stations.

The major domestic input, of course, is labor, and it, in turn, generates some indirect impacts on the Zambian economy. The government levies income tax on wages earned. These wage earnings also create demand for goods and services produced by the other sectors. Since the average wage rate in the economy (including mining) in 1974 was about 60 percent of the mining wage, mineworkers have a disproportionate share of the spending power.

The trend of wages in the mining industry seems to affect the general price level as well as wages in the rest of the economy. The effect on the rest of the economy seems to have been more important after independence. In 1966, following a report by the Brown Commission, the Zambian Mineworker's Union was granted a general wage increase of about 22 percent. Table 1–6 summarizes

Table 1-6
General Wage Increases after the Brown Report

Industry	*Commission*	*Date*	*Percentage Increase for Lowest Level*	*New Minimum Wage/ Month (Kwacha)*	*Total Labor Affected*
Mining	Brown	1 October 1966	22.0	54.60	43,500
Central government	Whelan	1 January 1967	13.5	27.00	35,800
Urban government (local)	Negotiation	1 January 1967	4.0	28.20	8,900
Rural government	Ministry of Local government	1 January 1967	20.0-80.0	27.00	2,400
Shopworkers	Wages Council	1 November 1966	21.0-25.0	26.20-30.00	5,900
Building engineering	Joint Industrial Council	1 April 1967	33.0	27.40	21,000
Hotels	Wages Council	1 May 1967	33.0-35.0	28.00-36.00	4,100
Agriculture	Wages Council	1 August 1967	29.6	14.40	30,000
General minimum wage	Wages Council	1 April 1967	33.0	27.40	17,000

Source: J.B. Knight, "Wages and Zambia's Economic Development," in Charles M. Elliot (ed.), *Con-Straints on the Economic Development of Zambia* (Nairobi: Oxford University Press, 1971), p. 103.

the subsequent wage increases in the rest of the economy.[15] This single round seems to be but one example of a prevalent trend.

The export earnings from copper have affected the price level through their impact on the money supply. Mineral revenues have a direct bearing on the overall level of government revenues and the financing of government expenditures. In times of low copper prices, the government could choose to borrow from the Central Bank and thus increase the money supply. When copper prices are high, the balance of payments improves, and generally, in the absence of countervailing policy, so does the money supply.

The benefits from mining appear to accrue to those Zambians who live in the urban areas and to increase the disparity in income between the urban and rural dwellers. This, in turn, leads to migration of the population from rural to urban centers, with economic and political consequences. The major transmission mechanism for mining benefits to rural areas is through government services such as schools, hospitals, and agricultural services. Children educated in rural schools, however, are likely to migrate to urban areas. Thus their education provides no lasting benefit to the rural areas. Agricultural services, including the provision of fertilizer to rural farmers, are aimed at improving productivity. Increased production would give the farmers surplus output to sell to the urban dwellers, and the ability of these urban dwellers to purchase the product is in part dependent on direct and indirect income from mining. Rural dwellers also benefit from income transfers from their relatives who are employed in mining and other urban sectors.

There are many other positive and negative effects. We have outlined only the ones we consider the most important. These effects are dynamic and can only be quantitatively analyzed by a simultaneous-equation model approach. However, it is impossible to capture all the effects. For instance, the effects of government policies toward the rural areas may be difficult to quantify because they are so diffuse. The transfer of income from urban to rural dwellers through family circles may be impossible to model econometrically because of the lack of data. These are some of the shortcomings of the econometric-model approach that the reader must consider when evaluating its results.

Notes

1. For a full account of the early history of the Northern Rhodesia mineral industry, see chapter 3 of Richard Hall, *Zambia* (New York: Praeger, 1965).

2. Sir Ronald L. Prain, "Statement by the Chairman," RST Group of Companies, Zambia, 30 June 1965.

3. "Towards Complete Independence," an Address by the president, Dr. K.D. Kaunda, Lusaka, 11 August 1969, p. 35.

4. *Copperbelt of Zambia Mining Industry Yearbook, 1965,* p. 19. The shortages were ameliorated through the use of "volunteers" and overtime.

5. "Explanatory Statement of RST to Shareholders," 6 August 1970, Appendix G.

6. Roan Selection Trust, "Report of the Chairman," June 1967.

7. Republic of Zambia, *Second National Development Plan, January 1972–December 1976.* Ministry of Planning and National Guidance, December 1971.

8. The estimate is in constant price terms.

9. Some other routes were used, such as Mombasa in Kenya, Beira in Mozambique through Malawi or through Botswana.

10. South African goods could not be imported through Dar es Salaam because of the Tanzania embargo on South Africa.

11. See Carmine Nappi, *Commodity Market Controls* (Lexington, Mass.: Lexington Books, D.C. Heath, 1979), chapter 7.

12. R.S. Pindyck, "The Cartelization of World Commodity Market," *American Economic Review* 69 (May 1979).

13. James Fry and Charles Harvey, "Copper and Zambia," in *Commodity Exports and African Economic Development,* edited by S. Pearson and J. Cownie (Lexington, Mass.: Lexington Books, D.C. Heath, 1974). Wages and salaries in mining are much higher than those in other sectors.

14. RCM borrowed on the international money market, but through the Bank of Zambia. By 1977, its outstanding loans totaled U.S.$60.8 million payable in foreign currency and 29.6 million kwacha from the Zambian government. NCCM has U.S.$339.2 million in short-and long-term foreign loans and 68.3 million kwacha payable to the Bank of Zambia.

15. This relationship cannot be captured in a functional form if changes in the nonmining sectors are made a function of changes in the mining wage. This is because of the irregularity of the lag with which wages elsewhere respond to wages in copper production. The literature points to the existence of the relationship, however. Therefore, absolute wage levels are used in this study.

Export Instability and the Zambian Economy, and an Overview of the Model of the Economy

Introduction

As has been discussed, the Zambian economy is dominated by a single export commodity with a highly volatile world market price. This volatility in copper prices is transmitted directly to the Zambian economy through export revenues. Copper prices also affect industry profits, on which government tax revenues and dividend earnings are based. Since the industry supplies the major share of total government revenues, any fluctuations in profits is mirrored in government receipts. The Zambian government plays a major role in the economy, especially since 1968, when it acquired an interest in the country's major industrial enterprises and government financing became a major source of investment funds.

The copper industry employs over 14 percent of the total Zambian wage factor. In addition, as table 1–4 showed, the industry is a major user of domestic-industry products. Payments for these domestic inputs, are relatively stable in the short run because the industry does not make significant adjustments to its level of operations in response to short-run price movements.

The fluctuation of copper prices was not a serious problem in the Zambian economy until around 1970 because, in the years between independence (1964) and 1970, government revenues and foreign-exchange earnings did not constrain the economic growth of Zambia. The major constraint was the availability of domestic nonfinancial resources, particularly manpower. Copper prices were high and the export revenues generated were more than enough to meet Zambia's imports; thus, foreign-exchange reserves accumulated. Revenues from the mining sector kept government revenues higher than could be spent effectively, given the constraints imposed by other factors. After 1970, however, a combination of lower copper prices and rising imports and government expenditures resulted in both government and balance-of-payments deficits in 1971, 1972, and 1975. Government recurrent spending was also a deficit in 1976. New government policies, such as import licensing, foreign-exchange control, and a greater use of foreign financing, were introduced to deal with the new reality.

Export-earnings instability can originate from quantity or from price instability. Zambian data on copper export revenues, prices, and volumes indicate that price instability is the dominant cause of instability of export revenues. During the period 1957–1976, the variance of the log of the copper export price around a constant growth-trend line was 77 percent of the trend-adjusted variance of the log of export revenues. For the same period, the variance of the log of export quantities, adjusted for trend, was 20 percent of the trend-adjusted variance of revenues. The covariance of the logs of price and quantity was positive.[1] Before we analyze the impact of copper-price fluctuation on the Zambian economy, we shall briefly examine the empirical evidence on the subject of export instability.

Empirical Studies of Export Instability

The measurement, effects, and causes of export instability have received considerable attention from economists. Much of the literature deals with empirical studies of the causes of export instability, its effect on economic growth, and the differences in instability of exports from developed countries (DCs) and less-developed countries (LDCs). Particular attention has been paid to the effect of instability on the economic growth of LDCs.[2] Most of the studies start with the hypothesis that export instability hinders economic growth. Instability of exports reduces the capacity of the LDC to import the capital goods needed for capacity expansion. Export instability also increases the cost of foreign capital and the size of foreign-exchange reserves needed for financial credibility. Instability of export earnings results in unstable national income, savings, and investments and thus lowers the efficiency of investment. The consequence of export instability may be frequent intervention by the government in import and foreign-exchange markets. This induces socially suboptimal behavior in economic agents and may sometimes produce the opposite of desired results. Import restrictions, for example, generally lead to socially suboptimal inventory accumulation. They may also lead to losses in domestic production, caused by shortages of needed imported inputs. Export instability might result in unanticipated inflation and raise the domestic price level because of rachet effect.

The empirical work on the impact of export instability has been based on either a cross-sectional, cross-country approach, or a single-country, simultaneous-equation model approach. Coppock (1962) and McBean (1966) use the cross-country approach, although McBean conducted single-country case studies. The cross-country approach has been criticized on the grounds that it ignores differences in economic structures, dynamics, and policies of LDCs and because of the measurement problems associated with the approach.[3] The single-country, simultaneous-equation approach uses econometric models of commodity and macroeconomy to study instability effects. An example of this approach is the study of Ghana and its cocoa industry by Acquah (1972). A criticism of the single-country approach is that the results cannot be generalized.

The studies using the cross-sectional approach test for statistical significance between export instability and measures of economic performance, including growth of gross domestic product (GDP), growth rate of investment, marginal efficiency of investment, change in domestic-price levels, and stability of gross domestic output and other macroeconomic variables. These studies utilize simple and rank correlation tests and multivariate regression analysis. Most studies do not find any stable and statistically significant relationship between export instability and measures of economic performance. The results appear to be very sensitive to the sample, the time period, and the measure of export instability.

Coppock (1962) and McBean (1966) found no significant relationship between export instability and growth rate of GDP, and Glezakos (1973) found a significant negative relationship. Knudsen and Parnes (1975) found a significant positive relationship, using a measure of export income based on the permanent-income theory of consumption, a measure that is conceptually different from the measures used by McBean, Coppock, and Glezakos.

McBean found a positive but insignificant relationship between the growth rate of investment and his measure of export instability by simple correlations. However, one of his multivariate regressions showed a significant positive relationship between export instability and growth rate of investment. Kenen and Voivodas (1972) duplicated this result using the same sample and period used by McBean, but when they tested for other periods or expanded the sample, they found no significant relationship. Knudsen and Parnes found that the ratio of investment to GDP is positively and significantly related to export instability by using multivariate regressions. However, the instability index used in their test is the average instability of income in the economy. When export instability was used, the significance level dropped to 10 percent.

Cross-sectional tests of the determinants of the LDC demand for international reserves have revealed a stable and significant positive relationship between the demand for reserves and export instability. Foreign-exchange reserves serve to cushion the fall in international receipts; thus, reserve holdings should be an increasing function of the instability of international payments. Several measures have been used for the instability of international payments. Kelly (1970) and Iyoha (1976) use the standard deviation and variance of exports, respectively, as proxies for the instability of payments. In both studies, the coefficient of export instability was positive and very significant. Eaton and Gersovitz (1980) estimated the demand for international reserves, with and without capital rationing, and found the coefficient of the standard deviation of exports to be positive and significant in both cases. These results confirm the hypothesis that export instability increases the demand for international reserves by LDCs.

Several studies have used the single-country, case-study approach to study export instability and price stabilization. Rangarajan and Sundararajan (1976) developed econometric models of eleven LDCs and simulated these models under stable and unstable export regimes. They found that the impact of instability was not the same in all countries, and in five countries, the growth rate declined with an increase in instability of exports. However, the models used by these authors were simple Keynesian models, which ignored supply considerations, explained only aggregate quantities, and did not have a commodity-export sector. On the basis of their results, Rangarajan and Sundararajan conclude that the usefulness of international schemes aimed at stabilizing primary product prices or export earnings of LDCs has to be examined for each country separately on the basis of an analysis of its economic structure.

Acquah (1972) studied the impact of the fluctuations in the price of cocoa on the economy of Ghana and concluded that stabilizing the price of cocoa—Ghana's principal export product—would result in a potential growth rate that is superior to that under export unstable prices. Lasaga (1979) found that stabilizing the price of copper reduced the level and the variability of the aggregate price level in the Chilean economy.

The inconclusive nature of the cross-sectional tests of the effects of export instability makes case studies of individual countries and commodities necessary. Even if cross-sectional studies provided strong evidence in any direction, it would still be useful to study each country separately in order to identify the channels through which the effects of instability are propogated and to provide clues to suitable policies that would counteract any negative effects or strengthen positive ones. A well-specified econometric model that had a detailed commodity-export sector and a disaggregated macromodel, with all the germane linkages to the commodity sector, would be useful in studying and testing for the effects of export instability and price stabilization.

Zambian Government Behavior and Policy Responses to Instability

The effect of export instability depends on how economic agents react to it. Zambian government actions can reduce or exacerbate instability effects on economic growth and welfare. The government has to ensure that, in the face of fluctuating revenues, its expenditures do not deviate drastically from its revenues and that expenditures on imports are consistent with export earnings and foreign-exchange reserves needed for financial credibility. Policies to maintain a budgetary balance include making adjustments to current and planned spending, changing the time path of investment projects, and the accumulation and depletion of reserve funds depending on the copper price cycle. They also include borrowing and lending at home and abroad. Policies for maintaining a healthy level of reserves include foreign-exchange controls and import restrictions through quantitative controls and tariffs. The timing of these policies is crucial to their effectiveness, and the timing, in turn, depends to a great extent on the ability of the government to forecast the trends in copper prices. If the timing is wrong, the policy may cause costly dislocations to the economy without contributing any benefits.

A possible response to fluctuations is for the government to make its plans consistent with the expected long-term trend level of prices. It can then borrow and lend domestically and abroad to smooth out short-term fluctuations. There are problems associated with this policy, however. First, there is the question of forecasting the long-term trend of copper prices. Second, there are costs associated with the policy. In a world where capital markets are imperfect, borrowing rates exceed lending rates because of transactions costs and risk aversion. In international financial markets, non-Zambian financial intermediaries are the beneficiaries of the transaction costs. Thus, borrowing and lending from external sources imply a transfer of wealth from Zambia to abroad. Borrowing and lending in the domestic economy may have consequences on the explicit or implicit interest rates, investment, and price stability. Government borrowing from commercial banks to finance deficits may have a crowding-out effect—reducing commercial bank lending to the private and parastatal sectors and reducing investment. Government lending from a budget surplus may increase bank liquidity and the money supply and hence the price level.

The aforementioned policy might be optimal if international and domestic capital markets were perfect. With imperfections in markets, Zambia can lend as much as it wants without affecting the lending rate, but it cannot borrow as much as it wants at the same rate of interest. Given that copper prices exhibit wide fluctuations, with long time cycles, the need for borrowing can grow very large, especially if adjustments are not made to imports or government expenditures. Large or long declines of copper prices may lead to defaults in servicing existing loans or to refinancing, and both are costly. Defaults on loans make reentry into the market difficult and more costly. As a result of the decline of copper prices in 1975–1977, Zambia borrowed heavily abroad, with the expectation that copper prices would rebound. By 1978, copper prices were still depressed, and Zambia had difficulties meeting its financial obligations. An attempt to secure foreign loans toward the end of 1978 could not find enough participants, even at the very attractive rates offered.

The application of government policy instruments to deal with copper price fluctuations increases the degree of uncertainty. Import quotas, tariffs, and degree of enforcement of existing controls change from time to time, depending on the level of foreign-exchange reserves. As a result, they may leave importers uncertain as to when and how much they will be allowed to import. Import controls can also cause delays in the importation of goods which are in existing industries and investment projects. The implications of this uncertainty to risk-averse individuals are (1) that their level of investment may be lower, (2) that the level of output may be lower, and (3) that the level of inventories may rise due to the uncertainty of lead times. Fluctuations in export revenues hinder economic progress to the extent that such policies have to be designed to counteract it. With regard to changes in spending in response to changing revenues, there are adjustment costs involved in moving from one level of spending to another. Projects whose planning entails substantial resource costs may be cancelled. However, if expenditures are to be increased, projects may be rushed to implementation without proper feasibility studies and selection procedures. Frequent changes in government plans and spending increase planning and coordination costs. In addition, frequent changes in the fiscal program may lead to uncertain inflation.

Zambian government-spending plans are based on the government's forecasts of copper prices and production. The available evidence on these forecasts indicates that they have been consistently off the mark. Jolly and Williams (1972) analyzed the forecasting record of the government with respect to revenues and expenditures. They found that, for the period 1957–1970, the government consistently underestimated revenues and expenditures. Revenues were underestimated because, although copper production was estimated accurately, copper prices were grossly underestimated. A possible explanation is that the estimates were adjusted for the risk of price fluctuations. According to Jolly and Williams: "The general approach has been to err on the side of safety—to estimate government revenue on the basis of fairly pessimistic assumptions of copper prices."[4] This behavior reduces the probability of a budget deficit. This preference was influenced in part by the conservative approach to budgeting by preceding colonial administrations. The conservative-estimation approach may also be an attempt to dampen the expectations of the public. This could reduce the demand on the budget by various government departments, wage demands by government workers, and demand for government services. This approach to dampening expectations is by no means optimal, since it can also dampen the expectations of private investors and thus reduce investment. In addition, if individual rationality is assumed, continued use of this tactic would have no effect in the long run.

Government behavior and policies engendered by fluctuating and unpredictable prices would seem to be detrimental to economic growth. The need to adjust spending plans frequently in order to avoid deficits and enforce import controls to maintain foreign reserves would increase uncertainty in the economy. The administrative machinery necessary for the implementation of policies designed to counteract fluctuations uses up manpower that could be put to productive use. This is important for Zambia because of its administrative manpower scarcity. Since this machinery is needed mostly at the low end of the copper price cycle, it may be almost idle when the price is on the high end. In addition, these policies may be implemented inefficiently, which not only reduces their effectiveness but causes further dislocations in the economy. This is a nontrivial issue in many developing nations, where qualified and experienced management personnel is scarce. These costs are in addition to the adjustments and transaction costs that were mentioned earlier.

If copper prices were more predictable, some of these costs would be reduced or possibly avoided. Government may still borrow from abroad, but since the variance of its income is reduced, ceteris paribus, the interest rates at which it borrows would be reduced, and its capacity to borrow would be enhanced. The frequency with which it goes to the money market may also be reduced; it may reap benefits from economies of scale in addition to being able to borrow when interest rates are more favorable. With stable prices, the necessity for the government to change its plans in accordance with the copper-prices cycle and the associated adjustment costs will be reduced. Government plans and actions would then be dictated by changing conditions in the domestic economy rather than by those in the world copper market.

Price Fluctuations and the Mining Industry

Fluctuating prices affect the behavior and performance of the mining industry and its contribution to the economic growth and welfare of Zambia. The most direct and noticeable impact of price instability is on the earnings of the companies, which tend to rise when prices rise, and the converse. Company profits are used to pay taxes and dividends and to finance investments and working capital. Until 1970, all dividends were paid to nonresidents, but since then, the Zambian government receives 51 percent of the dividends in addition to taxes. The shares of the foreign mining companies with equity holdings in the Zambian industry are traded on major stock exchanges in London and New York. The cost of capital to these firms is reflected by the riskiness of their business. Mining is a very capital-intensive industry, with high operating risk. The combination of high operating risk and earnings volatility increases the cost of capital to the industry.

A consequence of the high cost of capital is reduced investments. To the extent that price volatility raises the cost of capital, it reduces investment. As a result of high risk, there is a tradition in the mining industry of payment of high dividends. This typifies the behavior of Zambian mining companies. Dividend payments of the companies between 1964 and 1969 averaged 80 percent of profits. Since most investment in the Zambian mining industry is financed out of earnings (presumably as a result of the high cost of outside financing), the result of high dividends is low investments. The asymmetry in the dividend behavior of companies makes it such that retained earnings (and therefore investment) bear the greater impact of fluctuations in earnings. When prices are at their peak, profits are very high and far exceed the need for investment and working capital, and only a small proportion of profits can be retained for these purposes. The Zambian capital market is unable to absorb such substantial inflows, so the funds are sent abroad, mostly as dividends. When prices fall, dividends have to be maintained, and retained earnings and investment probably suffer. Alternatively, it could be that the investment program is set so that it can be implemented with a low level of profits, such that dividend payments do not suffer. Insofar as it is desirable to expand the output of Zambian copper, lower investments may have been detrimental to economic growth.

The takeover has not substantially changed the relevance of this analysis. The cost of capital is still affected by price volatility. Even if the government were to operate outside the bounds set by market discipline, its existing or prospective partner would still be constrained by the market. The possibility of the government undertaking any major mining venture on its own is very remote. The ability of the government or industry to raise funds and repay loans still depends on the level of copper prices. The cost of borrowing is likely to be influenced by price volatility.

Price Fluctuations and the Rest of the Economy

The rest of the economy is shielded somewhat from the direct impact of short-term price fluctuations on income, since copper income is shared by the government and the mining companies. However, fluctuations do influence their expectations on future income, employment, price level, government spending, and their ability to obtain their preferred mix of investment, intermediate, and consumption goods. These expectations, in turn, have an effect on consumption, saving and investment, and the composition of investment. When the price cycle is on the high side, it is more difficult for the government and the copper industry to resist demands for higher wages by workers or for the government to resist demands for more services. On the low side of the cycle, money wages are not reduced, as they are generally inflexible downward, and it is very difficult to reduce services. The net effect is that, in times of favorable prices and revenues, government expenditures go up, but they are not correspondingly reduced when revenues fall. The result is a government-spending pattern that could generate surpluses in times of high prices and large deficits in periods of low prices.

The effect of the uncertainty with regard to government policy on the production, inventory accumulation, and investment decisions by agents in the economy has been discussed in a previous section. Another effect could result from the high required return in the mining industry. Since mining is the dominant industry, it is likely that the high required returns demanded in the industry may affect required returns by domestic and foreign investors in other, less risky industries.

Price Fluctuations and Inflation

Fluctuations in the price of copper might affect economic growth and welfare through their effect on the domestic economy's price level. Inflationary pressures might develop through budget deficits and surpluses and increased demands for wages and services at times of favorable copper prices. Since the generation of budget deficits and surpluses is a result of the movement of copper prices, which is an event exogenous to Zambia, deficits by themselves are not inflationary and surpluses are not deflationary.[5] Inflationary pressures are generated by the methods of financing the deficit and the reaction of the government in changing its spending patterns in the event of a surplus. The general practice in financing deficits in Zambia is a combination of domestic and foreign borrowing and drawing down of government accounts in the central and commercial banks. The deficits that accumulated in the period 1971–1976 were financed as follows: about 61 percent from domestic borrowing, 29 percent from foreign loans and grants, and the rest by drawing down government accounts in central and commercial banks. For the same period, the Central Bank financed 54 percent of the domestic borrowing, the commercial banks 36 percent, and the rest was financed by the nonbank public.[6] The major instruments of government debt are treasury bills and short-term bonds.

Financing deficits through borrowing from the Central Bank and drawing on Central Bank accounts and through foreign loans and grants leads to monetary expansion, which may generate inflationary pressures. Borrowing from commercial banks should not normally lead to monetary expansion, although it can have a crowding-out effect, tightening credit and discouraging investment. However, Zambian treasury bills and government short-term bonds can be used by commercial banks to meet liquid-reserve requirements. Thus, borrowing from commercial banks can lead to monetary expansion.

Increases in the price of copper lead to increases in the value of exports, and thus to improvement in the balance of payments. Mining profits and therefore government revenues are also increased. The opposite effect obtains in the case of price declines. The increase in government revenue increases budget surplus in the short run. Under the takeover arrangements, by which only 49 percent of declared dividends go to foreign shareholders, an increase of foreign assets held by Zambian residents is assured. This increase has the effect of increasing the domestic money supply if the Central Bank does not take any action to sterilize it. However, there is also the opposite effect that, with a budget surplus, the need to borrow from the Central Bank is reduced. This analysis of the effects on monetary expansion is a static one and does not fully consider some secondary effects, such as the impact of increased or decreased exports on imports and the balance of payments, both of which have an effect on monetary expansion. These secondary effects can only be captured in a dynamic analysis.

Unless properly managed, fluctuating prices can lead to unanticipated inflation through their effect on the domestic money supply. This can have several consequences for Zambia. It widens the income differential between the rural and urban sectors, since the urban workers are better placed to protect themselves from inflation through wage bargaining. It may necessitate constant devaluations of the domestic currency, raising the price of imports, which in turn increases the inflationary pressures. Increases in import prices can hurt many of the existing industries that depend on imported inputs. It may well have an effect on domestic investment, which depends on imported capital. Zambia depends heavily on expatriate personnel in many of its industries. To be able to continue to attract skilled expatriates, who are consumers of imported goods and remit some of their earnings abroad, Zambia needs both price and exchange-rate stability.

Conclusions on the Effects of Export Instability

The foregoing analysis of the effect of copper-price instability would lead to the conclusion that instability affects the Zambian economy adversely by (1) the uncertainty of government and industry revenues, making it difficult for them to plan effectively and reducing their capacity to borrow abroad; (2) the uncertainty, costs, structural dislocations, increased demand on scarce resources, and administrative inefficiencies caused by government policies designed to counteract instability; and (3) the higher cost of capital to the mining industry, with implications for investments and dividend payments to nonresidents. These adverse effects are exacerbated by the dominating position of copper mining in the economy and in exports, the lack of or the thinness of the domestic capital markets, and the scarcity of qualified manpower. The adverse effects could be reduced if Zambian government and industry were able to forecast prices accurately. However, this would require additional investment in information collection and model building and would further strain the resource constraints.

One of the objectives of this book is to establish if and how the Zambian economy is adversely affected by price instability. This objective is to be accomplished by dynamic simulation experiments using an econometric model of the Zambian economy. However, some of the problems inherent in this approach must be pointed out in advance. First, it is generally agreed that, immediately after independence, export and government revenues were not constraints on economic development in Zambia. Nevertheless, we have to use data from this period in our model. Second, an econometric model is derived from data on past behavior of the economy, that is, behavior under certain conditions. In the case of Zambia, the model would describe behavior under fluctuating prices. Thus, to test the effect of instability adequately, we need another model of behavior under stable prices for comparison. Since these circumstances never existed, we lack the necessary data for them, and therefore no adequate model can be constructed. One can speculate, of course, on how a particular behavioral relationship might be affected if export prices were stable. Finally, there is the problem that confronts all quantitative work in economics—data. Not all effects and behavior are quantifiable, and even if they were, the data may not exist or may not be reliable; for example, administrative inefficiences are difficult to quantify.

In spite of these difficulties, we shall proceed to construct a model to test the effects of price instability. The hypothesis to be tested is that copper price instability is not detrimental to the economic growth of Zambia. The aforementioned difficulties should not invalidate the test. The simulations could provide useful insights into the dynamic processes that propagate instability. Using the test results and drawing from economic theory and prior empirical work, we can arrive at some useful conclusions. In the rest of this chapter and the chapters that follow, we shall describe the model of the economy and perform and report the results of the test.

The Model: Main Features and Linkages

The objective is to build an econometric model of the Zambian economy that can be used to analyze the direct and induced effects of changes in the world copper market on the Zambian economy. The model should allow us to say something about the effects of changes in the commodity market on domestic economic variables such as income, investments, unemployment, prices, and balance of payments. It sould also be possible to use the model to simulate various policies that can be used to countervail any negative effects of the fluctuations in copper prices.

The model features a detailed domestic commodity (copper) sector model (the micro model) embedded in a macroeconometric model (the macro model). There is an explicit two-way link between the micro and macro models. This approach to dynamic analysis of commodity market effects has seen several applications to other-country, other-commodity combinations. Two such applications are in Acquah (1972) for Ghana and cocoa, and Adams, Behrman, and Roldan (1979) for Brazil and coffee. These countries have had a greater degree of economic integration between the commodity sector and the macroeconomy than Zambia has had. The commodity sector in these cases is also not as dominating in the economy as it is in the Zambian case. Nevertheless, although Zambian actions will have little effect on the world market, this cannot be said of Brazil on the world coffee market or Ghana on the cocoa market. We are more comfortable in assuming that Zambian actions, within the realms of reasonable policy, have no significant effects on the world copper market.

The relationship between the Zambian economy and the export sector has been studied elsewhere, but nowhere has the model of the mining industry been explicitly integrated into a macroeconomic model. Baldwin (1966) looked at the backward linkages between copper production and the rest of the economy, using a production-function approach. The major input turned out to be local labor, and he focused on the historical supply conditions in what was essentially a dual economy with surplus labor. There was no empirical investigation of the structural relationship between world market conditions and the Zambian copper industry and the Zambian economy or of how the rents earned by the government were related to economic development in general. Nziramasanga (1973) concentrated on the relationship between the copper market and the Zambian industry but not on the general growth effects. Blitzer (1978) used the Zambian case to illustrate the relationship between development and income distribution in a dual economy. He used a partial-equilibrium, fixed-coefficient growth model of an economy producing three goods (agricultural, manufactured, and mining output), with the agricultural sector divided into subsistence and commercial farming. All goods were considered tradeable and input-output ratios were considered fixed in the export (mining) sector. The world price of copper affected only the net trade deficit, not economic growth or income.

The econometric model, simultaneous-equation approach that we propose to use should be an improvement on these other studies of the relationship between the export sector and the Zambian economy. In our model, the major links between the macro and micro models are (1) export revenues earned by sales of copper to foreign countries, which are used by all the agents in the economy for imports of capital and consumption goods; (2) government revenues from the commodity sector in the form of mineral and company taxes; and (3) demand for and payments to local factors of production by the commodity sector. These are the direct effects, and they have an impact on other variables, such as balance of payments, money supply and price level, consumption, and investment. The feedbacks to the commodity sector are in terms of costs and profits, investment and financing in the commodity sector, and government policies toward the sector.

The Data

The major sources of data for the macro model were the World Tables (1976) of the World Bank, the *Monthly Digest of Statistics* of the Central Statistics Office, Zambia, and the *International Financial Statistics* of the International Monetary Fund. The data for the commodity-sector model (micro model) were compiled from various issues of the financial and technical reports of the copper companies. These sources were augmented by data obtained from the *Zambian Mining Yearbook,* which is published annually by the mining industry.

Data can impose severe limitations in building an econometric model. In the case of Zambia, there are problems concerning both quality and quantity of data. Zambia has existed as a separate economic entity only since 1964, so the length of time-series data is limited. This rules out the use of complicated and long-lag structures in the estimation of some of the behavioral relationships in the macro model. It also restricts the number of independent variables, so the estimated relationship may have some degrees of freedom. In addition, there are several interesting variables for which no data are available or for which the data available are not of sufficient length to be used in estimation. For example, data on sectoral capital stocks and investment are available for too limited a period. Data on labor force and unemployment are not readily available. This lack of data limits the scope of the model and the analysis that can be done with it. In some cases, even the quality of available data may be suspect. Because of frequent revisions and changes in the coverage of individual time series, there can be serious problems of consistency. In particular, the basis of reporting a time series may have changed, but the data in periods prior to the changes are not adjusted to reflect the change. In these cases, we have had to make adjustments to reflect the change. Fortunately, the cases in which adjustments were needed were few, and the changes needed were not very substantial.

The data used in the model are in annual series, since this is the only kind available. The equations were estimated with data up to 1976. Since we began the study, there have been further revisions to some of the variables in our data set. We have refrained from revising the series so that we can have a consistent set. The data for the commodity sector have longer time series, usually going back to the early 1950s, and the quality also seems to be better. The preindependence data are usable for the commodity model because mining operations were not immediately affected by the emergence of Zambia as an independent nation. However, the reorganization of the industry in 1970 seems to have affected the level of disaggregation of data available for the copper sector. For example, prior to 1970, detailed data on the operation of individual mines were available; since reorganization, such data are only available on a division basis.

Modeling a Developing Country

Apart from problems of data quality and quantity, there are other problems in building an econometric model of a developing country. Behrman (1978) and Behrman and Hanson (1979) provide a general discussion of the problems of modeling LDCs. Our discussion here will focus on how some of these problems apply to Zambia. Most of the available and usable data concern the urban sector and the money economy. Thus, the economic activities of the majority of the Zambian population are taken as a residual, with the representation of the urban-rural interaction through the migration statistics. Often models of economic behavior that are developed for advanced countries are applied to LDCs even when they may not be appropriate. Also, some models of behavior applicable to one LDC may be applied to another LDC without consideration of their appropriateness. Even within a country there may be considerable differences between various sectors, depending on their ownership structure and degree of capital intensity. For example, in Zambia, output in the construction sector could be demand-determined, since there is a wide latitude for substitution of labor for capital. However, in the manufacturing sector, output would be supply-determined, since there is less scope for substitution of labor for capital and the country has no capital-goods industry. Almost all capital goods have to be imported, and, given the transportation constraints on imports, there will be a slow adjustment to a desired level of capital stocks. Thus, while the construction sector is Keynesian, the manufacturing sector behaves according to the classical model.

It is usually assumed in LDC models that there is a labor surplus and that an employer can get all the labor he needs at the going wage rate. There is thus an aggregate, infinitely elastic labor-supply schedule. It is thus implied that labor is not a constraint to development. In Zambia, however, skilled labor is a constraint. Although many people may be officially classified as unemployed or underemployed, there is a severe shortage of skilled personnel and people with sufficient formal education to be easily trainable for jobs in the modern urban sector. The shortage of skilled people imposes a constraint on the number of unskilled people who can be employed as well as a constraint on the growth of the economy. The existence of powerful unions raises the wages of the unskilled far above the labor income in the informal and subsistence sector. Most new employees have to be trained for their jobs, so there is high investment of resources in each new employee. Adjustment to the desired work force level may therefore be slow. At the low point of the economic cycle, employers may be reluctant to lay off employees they have trianed, since they may not get them back or get equivalently qualified people when the economy turns up. At the high point of the economic cycle, employers may not have enough resources left over to train as many workers as they need. The implication is that the assumption of short-run profit maximization may not be adequate; the persistence effect in the use of labor must be explicitly recognized.

Another problem in developing-country modeling is the classification of expenditure items either as investment or current consumption. The national accounts of LDCs follow the pattern of accounts used by the developed nations. For Zambia, expenditure of education is classified as current consumption. However, education in Zambia is as much an investment as an acquisition of plant and equipment. The speed with which Zambia expands its educational facilities and opportunities is crucial to the attainment of high rates of economic growth and a greater degree of economic independence. Development-economic theory favors saving as an important element in the attainment of high growth. A significant proportion of Zambian government current expenditure goes to education and health, which really are investments in human capital formation. Therefore, a model based on current national-accounts schema must be careful about which elements of government expenditure to discriminate against in favor of savings in recommending optimal economic growth strategies.

A properly constructed model of a developing country such as Zambia can be useful in several ways. It can be used to study both short-term stabilization and long-term growth policies. Although development-economic theorists have in the past stressed economic growth and neglected short-term stabilization, economic stability appears to have a strong relationship with political stability, especially in African countries. Other things being equal, political stability should promote economic growth. An econometric model can be a useful aid for economic planning. It can be used to simulate the effects of different policy proposals. There is the problem, of course, that, if the data are of questionable quality, a model based on the data set will also be of questionable quality. However, if the econometric model uses a data set with which eventual policy decisions are made, it is likely to be an improvement on the fixed-coefficient, partial-equilibrium growth models that are based on the same data set on which many policy decisions may be based. In addition, the process of building an econometric model is a great learning experience. It improves the knowledge of the structure and dynamics of the economy and could provide clues to the improvements to the economic structure that would be beneficial. Building and using an econometric model could be an incentive to improve the quality and quantity of data, which would lead to better decisions overall.

General Model Characteristics

We assume the existence of a dual economy.[7] The modern sector (including commercial agriculture) has a significant raw-commodity export industry that is a price taker. Money wages in the whole sector are rigid downward, and the supply of labor is a function of this money wage deflated by the price index of domestically produced goods. There is less than perfect substitutability in production, so noncompeting imported capital, intermediate inputs, and skilled labor are always necessary, although the latter could be reduced by exogenously determined training programs. There is a ceiling on interest rates, and financial institutions have to ration the available credit. The supply of money is endogenously determined, and exchange rates are fixed.

The other sector is the traditional sector, where the capital-labor ratio and the marginal product of labor are low. The resulting lower average income prompts migration from this sector to the modern sector, with the individual immigrant basing his decision on the urban-rural income differential.[8] Since nominal wages in the modern sector are inflexible downward, however, there exists a possibility of unemployment in the urban area that may not be eliminated by an increase in aggregate demand.[9]

The behavior of this model is shown in figure 2–1. Both supply and aggregate demand in the modern sector are a function of prices. An exogenous increase in the world demand for the export commodity would raise the market price. If there were excess capacity, this would result in a relatively quick increase in production. Should the industry be operating at full capacity, then an expected increase in demand (and prices) would cause an increase in investment activity, based on the planned change in capacity and output. Relative factor prices and the technical requirements would determine the resultant increase as well as the composition of the labor force. The foreign-exchange inflows from investment, domestic factor payments, and lagged incremental export earnings would increase the money supply, shifting aggregate demand upward from D' to D'' (figure 2–1a). There is an excess demand for domestically consumed goods at the old equilibrium price P_0, and this pushes prices up to P_1. Equilibrium output also increases Y_0 to Y_1, generating an increase in total employment (from N_{t_0} to N_{t_1} in figure 2–1c), depending on the nature of factor proportions in the growing industries. A certain proportion in N_{d_1} will be domestic (unskilled) labor (figure 2–1d). In the domestic labor market, the demand for labor shifts to the right. The equilibrium real wage goes from W_0/P_a to W_1/P_a, where P_a is the domestic price index for consumer

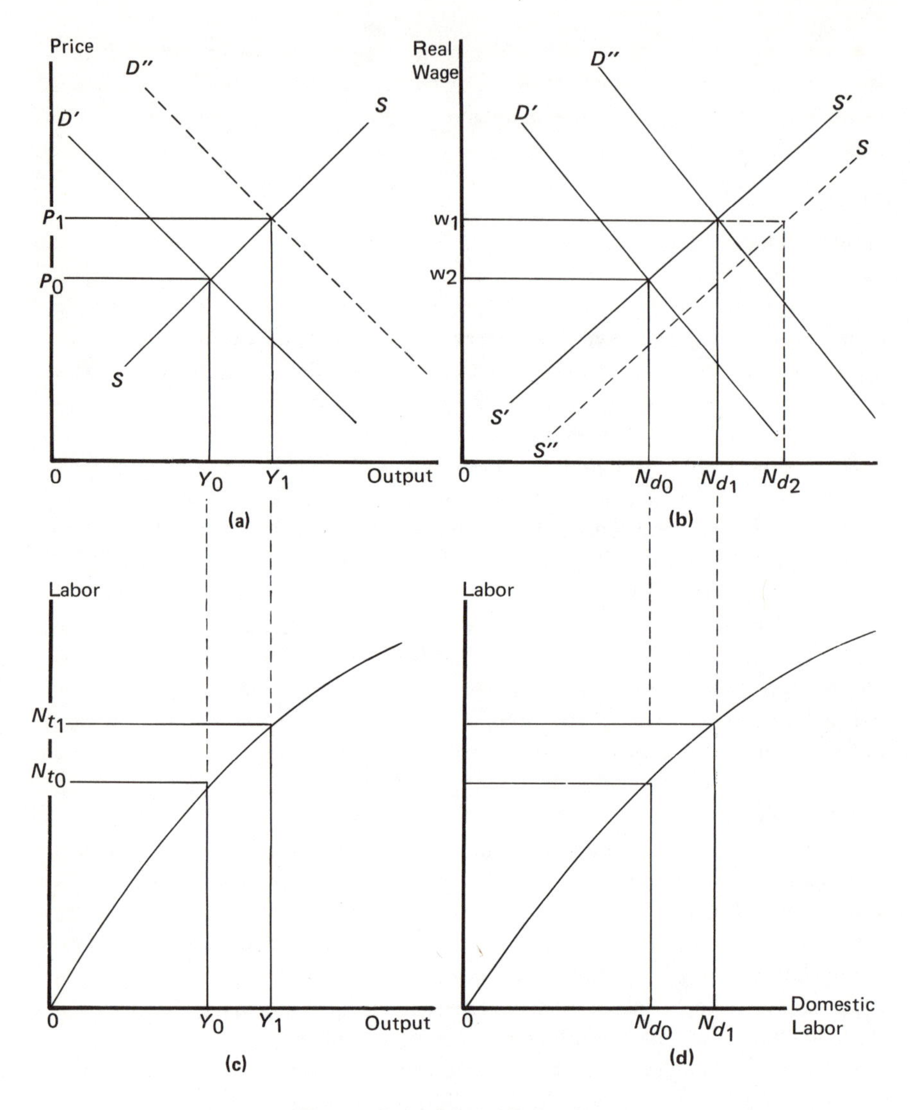

Figure 2-1. Model Behavior

goods (figure 2–1b). If P_a is relatively constant, this implies an increase in the urban-rural income differential, and an increase in migration. The labor-supply curve for the modern sector shifts to the right. Since the money wage is inflexible downward, however, this implies that urban unemployment (N_{d2}-N_{d1}) at the new exogenous change in the international commodity market would depend on the nature and length of the lags, the disposition of export revenues, and the size of the indirect linkage effects.

The purpose of the model is to establish the impact of instabilities in the international commodity market on the macro variables used to measure goal attainment in developing countries, such as employment and GDP growth rates. To do so requires the generation of a dynamic time path for the economy, assuming a smooth trend in either output or prices (or both), and then a comparison of the results with those obtained under unstable market conditions, where the instabilities involve random fluctuations around a long-term trend. A priori, we should expect random fluctuations in the impact variables (export prices, earnings) to produce results different from those of a smooth trend because of rigidities in the money wage, as has just been discussed. The presence of nonlinearities and lags in the functional forms or asymmetrical responses of certain variables to these fluctuations in export prices or earnings could have the same effect.[10] We could account for the monetary wage rigidity by imposing constraints on the response to negative changes in the independent variables. To test for asymmetrical responses of the components of aggregate demand to exogenous changes in the export sector, we included explicit measures of instability as independent variables. The measure tried was the annual deviation of prices (export earnings or foreign exchange) from their trend, divided by the trend value. To test for responses to price declines, this variable was set at zero whenever its value was zero or positive. Then it was reversed to test for responses to upward movements. The tests were applied to functions relating to government revenues and expenditures, public and private investment, and imports of intermediate, capital, and miscellaneous goods. None proved significant except imports of miscellaneous manufactured goods with respect to increases in export earnings. The model thus will look at the effect of secular trends in the export sector and their impact on the macro variables and how short-run fluctuations around this trend change the outcome, given the macro relationships, any rigidities in the money wage, nonlinearities imposed by the constraints, and the asymmetry of one of the import functions.

Notes

1. For a full exposition of the method of analysis, see Murray (1978).
2. The pioneering work in the area was by Coppock (1962), followed by McBean (1966), among others. Manger (1979) and Adams and Behrman (1980) provide critical and extensive reviews of the literature.
3. For example, see Adams and Behrman (1980).
4. See Jolly and Williams (1972).
5. See Harvey (1971, p. 44), and McBean (1966, p. 186).
6. These estimates were made using data from the *International Financial Statistics*, IMF, Sept. 1978.
7. See Behrman (1978) for a more complete discussion of the assumptions made here.

8. See Harris and Todaro (1970).

9. This type of unemployment can be reduced in the short run because of the lag between actual changes in the real wage and the formulation of new expectations by the prospective migrants about their urban wage. In the long run, however, only a reduction of the urban-rural income differential will reduce unemployment in the urban area.

10. For example, if the supply function is of the form $q = ln\, p$. The path of q when p is always on its mean value will be different from that obtained when p is fluctuating randomly around p. This type of nonlinearity was not suitable for any of our structural forms.

References

Acquah, P.A. 1972. *A Macroeconomic Analysis of Export Instability and Economic Growth: The Case of Ghana and the World Cocoa Market*. Ph.D. dissertation, University of Pennsylvania.

Adams, F.G., and Behrman, J.R. 1980. *The Commodity Problem and Goal Attainment in Developing Countries: An Integrated Econometric Investigation*. Lexington, Mass.: Lexington Books, D.C. Heath (forthcoming).

Adams, F.G., Behrman, J.R., and Roldan, R.A. 1979. "Measuring the Impact of Primary Commodity Fluctuations on Economic Development: Coffee and Brazil." *American Economic Review* 69:164–169.

Baldwin, R. 1966. *Economic Development and Export Growth: A Study of Northern Rhodesia 1920–1960*. Berkeley: University of California Press.

Behrman, J. 1978. "General Considerations of Model Specifications for Integrated Econometric Analysis of International Primary Commodity Market and Goal Attainment in Developing Countries." Unpublished manuscript, Wharton Econometric Forecasting Associates, Inc., Philadelphia.

Behrman, J., and Hanson, J. 1979. "The Use of Econometric Models in Developing Countries." In J. Behrman and J. Hanson (eds.), *Short Term Macroeconomic Policy in Latin America*. Cambridge, Mass: Ballinger.

Blitzer, C. 1978. "Development and Income Distribution in a Dual Economy: A Dynamic Simulation Model for Zambia." World Bank Staff Work Paper No. 292.

Coppock, J.D. 1962. *International Export Instability*. New York: McGraw Hill Book Co.

Eaton, J., and Gersovitz, M. 1980. "LDC Participation in International Financial Markets." *Journal of Development Economics* 7:3–21.

Glezakos, Constantine. 1973. "Export Instability and Economic Growth: A Statistical Verification." *Economic Development and Cultural Change* (July): 21:670–678.

Harris, J.R., and Todaro, M.P. 1970. "Migration, Unemployment and Development: A Two-Sector Analysis." *American Economic Review* 60:126–142.

Harvey, C. 1971. "The Control of Inflation in a Very Open Economy: Zambia 1964–9." *Eastern African Economic Review* 3:41–63.

Iyoha, M.A. 1976. "Demand for International Reserves in Less Developed Countries: A Distributed Lag Specification." *Review of Economics and Statistics* (August): 63:351–355.

Jolly, R., and Williams, M. 1972. "Macrobudget Policy in an Open Export Economy: Lessons from Zambian Experience." *East African Economic Review* 4:1–27.

Kelly, M.G. 1970. "The Demand for International Reserves." *American Economic Review* 60:655–667.

Kenen, P.B., and Voivodas, C. 1972. "Export Instability and Economic Growth." *Kyklos* 25:791–804.

Knudsen, O., and Parnes, A. 1975. *Trade Instability and Economic Development*. Lexington, Mass: Lexington Books, D.C. Heath.

Lasaga, Manuel. 1979. *An Econometric Analysis of the Impact of Copper on the Chilean Economy*. Ph.D. dissertation, University of Pennsylvania.

Manger, J. 1979. *A Review of the Literature on Causes, Effects and Other Aspects of Export Instability*. Unpublished manuscript, Wharton Econometric Forecasting Associates, Inc., Philadelphia.

McBean, A. 1966. *Export Instability and Economic Development*. Cambridge, Mass.: Harvard University Press.

Murray, David. 1978. "Export Earnings Instability: Price, Quantity, Supply, Demand." *Economic Development and Cultural Change* 27:61–73.

Nziramasanga, M. 1973. *The Copper Export Industry and The Zambian Economy*. Ph.D. Dissertation, Stanford University.

Rangarajan, C., and Sundararajan, V. 1976. "Impact of Export Fluctuations on Income—a Cross Country Analysis." *Review of Economics and Statistics* 58:368–372.

The Copper-Mining-Sector Model

The General Model

The copper-sector model provides the link between the Zambian economy and the world copper market. The equations in it explain copper production, the realized export price, the demand, and payments to factors of production. It assumes that actions by the Zambian copper industry have no effect on the world copper market.

We assume the industry has as its objective the maximization of profits. Other objectives could be considered relevant, especially since 1970. Until then, profit maximization clearly was the goal, with the government maximizing tax revenues, foreign-exchange earnings, and employment within the context of profit maximization. From 1971 to 1974, the takeover arrangements of 1971 and the management contracts discussed in chapter 1 were in effect.[1] They explicitly stipulated profit maximization as the goal of the industry,[2] and this of course soon conflicted with the Zambian government's intention to maximize domestic employment and production. Our strategy is to estimate the model, using profit maximization as the goal, but to provide the capability to simulate the model under other assumptions, such as production and employment maintenance and growth.

Copper Production

Profit maximization implies a supply function, with world copper prices and Zambian unit-production costs among the independent variables. The model uses the familiar stock-adjustment model.[3] This specification recognizes that, in the mining industry, adjustment of actual to desired output in response to price movement is slow. It employs a Koyck lag mechanism to show supply as a function of weighted present and past real prices of copper, with the weights declining geometrically with the length of the lag.[4] In order to reduce the effects of heteroscedasticity, we formulate the relationships in log-linear form. The desired supply (S_t^*) at time t is given by

$$S_t^* = \alpha + \beta \log P_t \tag{3.1}$$

where P_t represents real copper prices. The difference between the desired and actual production (S_t) is eliminated over a long period, that is,

$$\log S_t - \log S_{t-1} = \mu (\log S_t^* - S_{t-1}) \tag{3.2}$$

where μ represents the speed of adjustment of actual to desired supply. Combining equations 3.1 and 3.2, we obtain

$$\log S_t = \mu\alpha + \mu\beta \log P_t + (1 - \mu) \log S_{t-1} \tag{3.3}$$

The short-run price elasticity at the mean is given by the coefficient of the price variable, with the long-run elasticity computed as the short-run elasticity divided by the coefficient of adjustment (μ). We assume exports and the realized price to be linear functions of production and the London Metal Exchange price (P_t), respectively.

Links to the Zambian Economy

The product of copper exports and the realized price equals the export revenues (R). Government revenues from copper (G_m) are then derived from industry profits (π) as follows:

$$G_m = \Phi (R_t - C_t - T) \tag{3.4}$$

where Φ represents the tax rate on profits, C_t is production costs, and T is transportation charges. Government revenues from copper are determined by the tax laws and applicable tax rates for each period. In 1973, the effective tax rate was 73.05 percent of profits, less allowances for capital expenditures. It has since increased.

The other links to the domestic economy originate from the demand for inputs into the production process and investment into the industry. Production costs are divided into payments for labor and intermediate inputs. The demand for labor per unit of output (L/Q) depends on the capital per unit of output (K/Q) and the quality of the copper ore being mined (g); the lower the grade, given a level of refined output, the higher the demand for labor. The demand for intermediate inputs (I/Q) depends on similar factors as well as the quantity and location of the ore reserves at the beginning of the year (M).

These relationships are:

$$\frac{C}{Q} = \frac{WL}{Q} + \frac{I}{Q} \tag{3.5}$$

$$\frac{L}{Q} = (g, \frac{K}{Q}), \qquad g = \bar{g} \tag{3.6}$$

$$\frac{I}{Q} = i\,(g, K, Q, M), \qquad M = M_{-1} - Q_{-1} \tag{3.7}$$

$$w = \Theta\,(\frac{L}{Q}, \pi_{-1}) \tag{3.8}$$

$$K_i = k\,(P_i \ldots P_j),\; P = P_{\mathrm{LME}} \tag{3.9}$$

Equation 3.9 links the average annual wage for local labor to labor productivity, measured in man-years per unit of output (L/Q) and lagged profits; this takes into account the notion of price expectations as well as the concept of the ability to pay, often used by the Zambian labor unions in wage negotiations. The expatriate wage is explained only by productivity measured in man-years per ton of refined copper, because it is not unionized. Zambian copper mines use either open-pit or underground methods. Where both mining techniques are used, one method predominates. Equations 3.6, 3.7, and 3.9 were therefore estimated for both open-pit and underground mines.[5] Employment data for the industry are given in man-years and are classified as either of Zambian origin (mostly unskilled or semiskilled) or expatriate. The same breakdown is applied to wages per man-year. Equations 3.6 and 3.8 were therefore estimated by origin of the labor force. In open-pit mining, capital consists mostly of earth-moving equipment and is a substitute for labor. More mechanical methods for the removal of overburden, ore loading, and the like, means less labor per unit of output. In addition, the lower the grade, the greater the level of mechanization. In underground mines, a substantial part of the investment is complementary to labor. This includes tunnels for ventilation, transportation of labor, and so forth. The relationship between the level of investment and the labor-output ratio should still be negative but of a lower intensity than in open-pit mines.

Estimated Equations

Supply

The supply equation, equation 3.1, was estimated for Zambia for the period 1960–1976 to give

$$
\begin{aligned}
\text{LCUQCZM} = {} & \underset{(1.95)}{1.3720} + \underset{(6.01)}{0.7197}\ \text{LCUQCZM}(-1) \\
& + \underset{(1.84)}{0.0698}\ \text{LPRICE/COST} - \underset{(-2.52)}{0.0719}\ \text{DMUFULIRA 70/1} \\
& - \underset{(-1.79)}{0.0689}\ \text{DSZM6202} - \underset{(-3.95)}{0.1561}\ \text{DZ66}
\end{aligned}
\qquad (3.10)
$$

$$R^{-2} = 0.832 \qquad SEE = 0.0349 \qquad DW = 1.99$$

where LCUQCZM is the log of copper output measured as the copper content of concentrates, LPRICE/COST is the log of the LME price of copper in pounds sterling per metric ton, deflated by the Zambian cost index, consisting of a weighted sum of domestic and expatriate mining labor costs. The dummy variables represent the long labor strike in 1962 (DSZM6202), the unilateral declaration of independence by Rhodesia in 1966 (DZ66), the effects of which were discussed in chapter 1, and the Mufulira mine disaster of 1970 (DMUFULIRA 70/1), which reduced production in 1970 and 1971. The short-run price elasticity is very low (0.07). Although the long-run elasticity is much higher (0.25), it still reflects an inelastic response of production to price. The results also mirror the long upward trend in Zambian production during 1960–1970. Favorable prices, relatively low production costs because of high ore grades, and low wages for local labor combined to make the industry very profitable, resulting in an almost constant upward trend. An alternative specification of the supply function was used in the model simulations, and it assumed that a nationalized industry would opt for maximization of output. This would imply optimizing output regardless of the price level.

The Optimum-Output Specification

The first step is to estimate a relationship for the capital-labor ratio. We postulate that the log of the capital-labor ratio (LKKLR) is a function of a time trend (TIME), a measure of the capacity-utilization rate (CUR), and the lagged value of the dependent variable. With full utilization of capacity (CUR = 1.0), we get a value for the optimum capital-labor ratio. In the short run, the stock of capital is fixed; hence, the labor required can be calculated from the capital-labor ratio. The estimated relationship for LKKLR for the period 1961–1976 is

$$\begin{aligned} \text{LKKLR} = {} & \underset{(2.26)}{1.5379} + \underset{(3.27)}{0.0327}\ \text{TIME} - \underset{(-1.62)}{1.1058}\ \text{CUR} \\ & + \underset{(2.74)}{0.4423}\ \text{LKKLR}(-1) - \underset{(-3.01)}{0.3011}\ \text{DZ67} \end{aligned} \tag{3.11}$$

$$R^{-2} = 0.913 \qquad SEE = 0.0932 \qquad DW = 1.83$$

Then the employment (VTLE) can be calculated as follows:

$$\text{VTLE} = \text{VGOREAL}(-1) \,/\, \text{EXP}(\text{LKKLR}) \tag{3.12}$$

where VGOREAL (-1) is the lagged value of mining capital stock, measured in millions of kwacha.

The second step is to estimate a relationship for the output-labor ratio. The model assumes output to be determined by a log linear-production function. The log of the output-labor ratio (LOLR) is a function of the grade of ore milled (VOGRADE), the capacity-utilization rate (CUR), the log of the utilized capital-labor ratio (LKLR), and the lagged dependent variable. To recognize different rates of utilization of capital stock, the actual capital-stock series is adjusted by a measure of the degree of capacity utilization.[6] LKLR is thus obtained as

$$LKLR = \log(CUR * VGOREAL(-1)/VTLE) \tag{3.13}$$

where VGOREAL (−1) is the capital stock lagged one period. The capacity utilization rate (CUR) is obtained by a log linear interpolation of output peaks.[7] The capacity-utilization rate is also used as an explanatory variable for labor productivity.

The estimated result for 1955–1976 is

$$\begin{aligned} LOLR = & -0.5436 + 0.0474\ VOGRADE + 0.1141\ LKLR \\ & \quad (-1.84) \quad (2.03) \qquad\qquad (2.15) \\ & + 0.7108\ LOLR(-1) + 1.0788\ CUR - 0.0984\ DZ70/76 \\ & \quad (9.04) \qquad\qquad (3.91) \qquad (-2.57) \\ & + 0.2016\ DZ58 \\ & \quad (3.29) \end{aligned} \tag{3.14}$$

$$R^{-2} = 0.918 \qquad SEE = 0.0466 \qquad DW = 2.26$$

The dummy variable DZ70/76 represents the inclusion of the centralized services staff of the mining companies into the employment statistics from 1970. This also represents the posttakeover period. The dummy variable DZ58 represents an outlying employment statistic for the year 1958. The estimated relationship shows that the labor productivity is positively related to capacity utilization and the capital-labor ratio. One deficiency of this estimate is that the employment statistics are in man-years, not man-hours. There is no independent estimate of the intensity of labor utilization. The presence of the capacity utilization on the right-hand side serves as a proxy for labor utilization.

Given a level of employment from the capital-labor ratio estimate, the level of output is obtained from the following relationship:

$$CUQCZM = VTLE * EXP\ (LOLR) \tag{3.15}$$

With full-capacity utilization (CUR = 1), this could also be a measure of capacity output (CUQCZMP). When the profit-maximizing supply function is used in a simulation, the capacity-utilization rate could then be calculated as

$$CUQCZM\ /\ CUQCZMP \tag{3.16}$$

Employment of Labor

Copper-mining labor consists of Zambians and expatriates, the latter usually being skilled and professional people who are on fixed-term renewable contracts. Government manpower policy is to encourage Zambianization of employment in the mining industry, particularly in the more responsible decision-making positions. Employment relations are estimated for each of the labor groups. The equations estimated for Zambians are used in simulations where the profit-maximizing output equation is used. The equation for expatriates is used for all simulations. It is thus assumed that, in all cases, the ruling policy is to minimize expatriate employment.

The demand for local labor in open-pit mining, expressed as the labor in man-years per unit of finished production (VLORO), is a function of the grade of ore going through the concentrating mill (VGO), the value of operating capital in millions of kwacha (VGOCO), and its own lagged value. For the period 1954–1976,

$$\underset{}{\text{VLORO}} = \underset{(3.02)}{122.38} - \underset{(-2.10)}{28.2728}\ \text{VGOCO} - \underset{(-1.32)}{5.5969}\ \text{VGO}$$

$$+ \underset{(2.11)}{0.4826}\ \text{VLORO}(-1) - \underset{(-6.12)}{54.8222}\ \text{DZ51/69} \qquad (3.17)$$

$$R^{-2} = 0.84 \qquad SEE = 5.70 \qquad DW = 1.92$$

The dummy variable DZ51/69, equal to 1.0 for the period 1951–1969 and zero elsewhere, represents institutional changes after 1964 in the official attitude toward local mining labor. Such changes included the integration of the mining-union leadership into the government's wage-control boards as well as a general feeling that African labor had been underpaid as compared to expatriates prior to independence. This resulted in higher than normal wage-rate increases after 1964. The ore grade was not a significant variable in explaining employment of local labor in underground mining (VLORU). For the period 1954–1976,

$$\text{VLORU} = \underset{(0.19)}{2.2579} + \underset{(0.32)}{3.3770}\ \text{VGOCU} + \underset{(9.15)}{0.937}\ \text{VLORU}(-1)$$

$$R^{-2} = 0.80 \qquad SEE = 8.39 \qquad DW = 1.95 \qquad (3.18)$$

where VGOCU is the level of capital stock in underground mines, in millions of kwacha. As can be seen, the relationship between labor and capital is not as strong as in open-pit mining, and the t-statistic is not very significant. The positive sign seems to indicate that, for a given level of output, an increase in employment requires an increase in the capital stock, as had been anticipated.

The employment of expatriate labor (VTEE) used total labor from both types of mining and is a function of their real wage rate per year (WGAMQN), the total output of finished copper in thousands of metric tons (VTOTP), and a time trend. For the period 1954–1976,

$$\begin{aligned} VTEE = {} & \underset{(7.30)}{7352.92} - \underset{(-2.18)}{0.5873}\ WGAMQN + \underset{(1.91)}{4.9705}\ VTOTP \\ & - \underset{(-2.62)}{161.322}\ TIME \end{aligned} \tag{3.19}$$

$$\bar{R}^2 = 0.78 \qquad SEE = 693.14 \qquad DW = 0.61$$

The time trend accounts for changes in productivity and for the exogenous substitution of local for expatriate labor (Zambianization). Contrary to our expectation, the relationship is a demand rather than a supply function, with the real-wage variable being negative. The reason may be that the wage rate is not a good proxy for the actual compensation of expatriates. In addition to cash wages, they receive health, housing, and moving allowances, as well as lump-sum bonuses payable in foreign exchange at the expiration of their contracts. These additional payments are designed to counteract the skilled-labor shortage; unfortunately, we could not find any statistical data on them.

Total employment in mining (VTCE) is then a sum of local and expatriate labor:

$$VTCE = VLORO * VXTO + VLORU * VXTU + VTEE * VTOTP \tag{3.20}$$

where VXTO and VXTU are total output from open-pit and underground mining, respectively.

Wages

The average annual earnings for Zambian labor consist of the basic wage, overtime pay, and cost-of-living allowance. The payment of a bonus based on the profitability of the industry was discontinued in 1964, but unions have continued to use the ability-to-pay concept in their bargaining. Equations were estimated for wages paid to Zambians and expatriates. The real wage paid to Zambians (WGAMQZ/PC) is a function of the labor-output ratio (VTLE/VTOTP) and the ability to pay, measured by the lagged real gross profits of the industry (VGP/PC). The real wage paid to expatriates (WGAMQN/VPCE) is a function of the expatriate labor-output ratio (VTEE/VTOTP), the exchange rate between the Zambian kwacha and the pound sterling (EXCH), and the lagged dependent variable.

The estimated equations for the period 1957–1976 were

$$\text{WGAMQZ/PC} = \underset{(9.81)}{30.4814} - \underset{(-5.08)}{0.2069}\ \text{VTLE/VTOTP} + \underset{(2.83)}{0.6983}\ \text{VGP/PC}(-1) - \underset{(-7.15)}{5.7329}\ \text{DZF} \quad (3.21)$$

$\bar{R}^2 = 0.856 \quad SEE = 1.7208 \quad DW = 1.57$

$$\text{WGAMQN/VPCE} = \underset{(3.83)}{113.5376} - \underset{(-2.41)}{87.5223}\ \text{EXCH} - \underset{(-3.57)}{2.1913}\ \text{VTEE/VTOTP} + \underset{(2.85)}{0.3991}\ \text{WGAMQN/VPCE}(-1) - \underset{(-3.12)}{22.4812}\ \text{DZ72} - \underset{(-2.17)}{11.8509}\ \text{DZ70/76} \quad (3.22)$$

$\bar{R}^2 = 0.827 \quad SEE = 6.0098 \quad DW = 2.43$

The deflators of the nominal expatriate and Zambian wage (VPCE, PC) are the consumer price indexes for expatriates and Africans, respectively. They are determined in the macro model. The real profit is obtained by deflating nominal gross profits by PC. The dummy variable DZF was set to unity for 1956–1965 to represent preindependence conditions, and it shows that political conditions then militated against wage increases not justified by productivity increases. Pressures on employers and government for improved and equitable wages for African employees after independence led to the establishment of the Brown Commission in 1966, which awarded the African employees an average 22 percent wage increase. The same dummy variable DZF was not significant for expatriate wages.

The wage elasticities with respect to labor-output ratio were −0.94 and −0.35 for Zambians and expatriates, respectively. The long-run wage elasticity with respect to labor-output ratio for expatriates was −0.6. The short-run elasticity with respect to exchange-rate changes is −0.77. This means that, if Zambia devalues her currency, expatriate employees in the mining industry would expect their compensation to be increased. The dummy variable DZ72 represents the change in the classification of employees from African and non-African to Zambian and non-Zambian in 1972. Although this had little effect on the employment and average compensation of Zambians, it lowered the measured average compensation of the group formerly classified as non-African but now as non-Zambian. The dummy variable DZ70/76 represents the postnationalization period, which by its negative sign indicates that the companies were able to keep expatriate wages down. However, this may imply a shift in the method of compensation of expatriates—away from wages and salaries to higher allowances and increased lump-sum payments for contract employees. It may also reflect the effect of long-serving, high-salaried, permanent expatriate employees who decided on early retirement following the nationalization.

Demand for Intermediate Inputs

The value of intermediate inputs was obtained from the historical total-cost data by subtracting total wage costs from total operating costs (VTOPC). Real intermediate-input costs were then obtained by deflating with an index of imports of crude materials (PXD24G).

The equation is estimated for both underground and open-pit mining. The demand for intermediate inputs in underground mines, expressed in real kwacha per ton of finished production (VUMCU) for the period 1954–1976 is

$$\begin{aligned} VUMCU = {} & \underset{(0.173)}{22.943} + \underset{(0.25)}{9.4673}\ VGOCU - \underset{(-1.93)}{53.415}\ VGU * \underset{(3.70)}{1.2234}\ VXRCM \\ & + \underset{(7.08)}{179.150}\ DZ67/69 + \underset{(2.02)}{120.159}\ DZ51/69 \qquad (3.23) \\ & R^{-2} = 0.89 \quad SEE = 34.187 \quad DW = 1.82 \end{aligned}$$

where VXRCM is the output from underground mines. The dummy variable DZ67/69 represents disruptions caused by the unilateral declaration of independence in Rhodesia, discussed in chapter 1. The other dummy variable, DZ51/69, represents the prenationalization period, during which different accounting procedures were employed. The equation for inputs into open-pit mining (VUMCO) for 1965–1976 is

$$\begin{aligned} VUMCO = {} & \underset{(4.45)}{282.074} - \underset{(-1.32)}{27.357}\ VGOCO - \underset{(-2.23)}{23.3920}\ VGO \\ & - \underset{(-2.02)}{0.2562}\ VXNCCM + \underset{(4.96)}{75.830}\ DZ67/69 + \underset{(2.38)}{46.4837}\ DZ72 \\ & - \underset{(-3.16)}{64.1365}\ DZ74 \qquad (3.24) \\ & R^{-2} = 0.82 \quad SEE = 17.54 \quad DW = 1.87 \end{aligned}$$

where VGOCO and VXNCCM are the total capital invested in the output received from open-pit mines, respectively. The unilateral declaration of independence by Rhodesia (DZ67/69) adversely affected production costs. The dummy variable DZ72 represents the introduction of a new process, the leaching plant, for the treatment of copper tailings in the dumps.

Total intermediate input costs are the sum of the two

$$VIMNC = VUMCO * VXNCM + VUMCU * VXRCM \quad (3.25)$$

Sales of Copper

There are no data on refined copper inventories. We tried to test for the existence of some rational policy of inventory holding, without success, and we concluded that the difference between production and sales consisted of pipeline stocks. The sales of copper in thousands of metric tons (VTESP) were related to the finished output of copper (VTOTP), and the equation obtained for the period 1954–1976 was

$$\underset{}{VTESP} = \underset{(152.85)}{0.9952\ VTOTP} + \underset{(3.82)}{71.1805\ DZ69}$$

$$R^{-2} = 0.832 \qquad SEE = 17.97 \qquad DW = 2.308 \quad (3.26)$$

where DZ69 is a dummy variable for transportation problems. From 1966, after Rhodesia declared its independence from Britain unilaterally, the amount of copper exported through Rhodesia to the port of Beira in Mozambique declined and then ceased. Alternative routes had to be found. Exports and refined output did not normalize until 1969. Hence, in 1969 there was higher-than-usual sales activity. Concentrates that were stockpiled in 1966–1968 were refined, and pipeline stocks were depleted. However, even after 1969, the alternative transportation routes through Zaire and Tanzania were beset with temporary capacity bottlenecks at various times. Dummy variables were introduced to capture these effects, but they were insignificant. The reason probably lies in the model's use of annual data and the sporadic nature of the problems. Emergency measures allowed for the eventual exportation of annual production.

Realized Price of Sales of Copper

Zambian copper has traditionally been sold at contract prices based on the London Metal Exchange (LME) price. The realized price, obtained by dividing sales revenues by copper sales volume, is not equal to the LME price; however, it is closely related to it. It was therefore estimated as a function of the LME price converted to kwacha per metric ton (VZPRICE). The estimated equation for the realized price (VKZPRI) for the period 1954–1976 is

$$VKZPRI = -20.2045 + 0.9483\ VZPRICE - 267.0347\ DZ64/66 + 99.1553\ DZ69 + 52.1588\ DZ70/76$$
$$(-1.04)\quad (29.24)\quad (-13.61)\quad (3.25)\quad (2.92)$$

$$\bar{R}^2 = 0.991 \qquad SEE = 25.313 \qquad DW = 2.44 \tag{3.27}$$

VKZPRI is in kwacha per metric ton. The dummy variable DZ64/66 has a value of unity for the period 1964–1966 and is zero elsewhere. It represents the use of an administered producer price rather than a price based on the LME price during that period. The dummy variable DZ69 probably reflects favorable timing of sales in the year 1969, when demand was strong and prices were high. It may also reflect the dampening effect of strong product demand on sales discounts. The dummy variable DZ70/76 takes a value of unity for periods after 1970, the year of the nationalization of the industry, and reflects a change in accounting procedures. The elasticity of the LME price was slightly above unity, implying that world market fluctuations were directly transmitted to the realized price.

Revenues and Profits

Having obtained sales in physical volume and realized prices, we have total revenues (VSXCOP) as the product of sales (VTESP) and the realized price (VKZPRI):

$$VSXCOP = VTESP * VKZPRI \tag{3.28}$$

From employment and wages and intermediate inputs, we get total costs:

$$VTOPC = TLC + VIMNC + VTTCOST \tag{3.29}$$

where TLC is total labor cost, VIMNC is total value of intermediate inputs, and VTTCOST is total transportation costs. The unit transportation is taken as exogenous. Rail-transportation rates to the port have traditionally been subject to government negotiation; we assume ocean freight rates as given. From equations 3.28 and 3.29, we obtain an expression for gross profits:

$$VGP = VSXCOP - VTOPC \tag{3.30}$$

Net profits or after-tax profits are given by:

$$VGPNET = VGP - CGMR \tag{3.31}$$

where CGMR is part of the profits paid to the government in corporate and mineral taxes. CGMR is determined in the macro model. Part of the net profits goes into retained earnings, which is used to finance investment and working capital. The rest is paid as dividends, with 49 percent going to foreign shareholders and 51 percent to the Zambian government since 1970. Before nationalization, all declared dividends were sent abroad.

The contribution of the copper sector to the macroeconomy's gross domestic product is obtained as a function of the variable VXCOP, which is defined by

$$VXCOP = VSXCOP - VIMNC - VTTCOST \qquad (3.32)$$

Investment

Mining-industry investment consists of expenditures on the acquisition and development of mining properties; the acquisition of plant and equipment; and the development of necessary infrastructure, such as mill and concentrator, smelting, and refining facilities, power plants, maintenance shops, offices, and employee housing and services. For underground mines, capital expenditures are spent on sinking shafts, opening drifts, and installing elevators; for open-pit mines, they are spent on the removal of overburden. Investments in the industry are usually planned well in advance, and their implementation depends on the market outlook and the availability of financing. Major development projects usually stretch out over two to three years, sometimes more.

The decision to invest and the factors influencing investment depend on the objectives of the industry. The private shareholders in the industry would prefer an investment strategy that maximized profits. The government would prefer a strategy that maximized output, employment, and overall government revenues. There are therefore multiple objectives for an investment program to meet. For the private shareholders, investment should be based on the present value of future cash flow. These cash flows are very much dependent on future copper prices and government tax policies. The discount rate or the cost of capital should be high, reflecting the high risk of the industry. Financing considerations also play an important role. Investment to improve or expand facilities in ongoing mining operations is usually financed out of retained earnings. New equity and debt financing are reserved for new mining developments. Once the mine is established, it is maintained and expanded by internal funds.

In Zambia, almost all investment has been financed out of retained earnings since 1958. The companies carried very little long-term debt. The debt-equity ratio of the Roan Consolidated Mines (RCM) was 6 percent in 1970 and 14 percent in 1972.[8] Most of the loans since 1972 have come from the international money market, either through the Bank of Zambia (for the Roan Consolidated

Mines Division), or through foreign government-sponsored export-credit institutions (for Nchanga Consolidated Copper Mines).

The Zambian government is committed to the expansion of production and exploration and attempts to accomplish this by economic incentives to the mining and exploration companies and by its role as a majority shareholder of the industry. For example, between 1970 and 1973, the companies could write off investment expenditures against profits during the year they were made. Exploration expenditures could be written off against personal income if incurred by a Zambian resident. The government could expand investment by diverting its own share of dividends or tax revenues to mining investment. However, the government faces a financing constraint, too, because its ability to commit revenues to mining depends on the level of its overall revenues. The amount of revenue that the government receives is highly dependent on the level of copper prices.

For both government and industry, the level of copper prices and the expected future spot prices of copper are important considerations in the investment decision. Even if outside debt financing were to be used extensively, the cost of this financing and consequently the level of investment would depend on the expected future prices of copper, which determine the ability of government and industry to repay the debts. We therefore present two models of the investment decision. The first assumes that nominal investment is a function of past prices; the assumption here is that expectations of future prices are based on past prices. Investment in mining (KCMIN) is estimated as a polynominal distributed function (Almon lag) of past LME copper prices (PCLME). The resulting equation for the period 1964–1976 is

$$
\begin{aligned}
\text{KCMIN} = &-\underset{(-0.47)}{9.3544} - \underset{(-8.56)}{68.1}\ \text{DZ69/70} + \underset{(0.61)}{0.0119}\ \text{PCLME} \\
&+ \underset{(1.88)}{0.0262}\ \text{PCLME}(-1) + \underset{(0.71)}{0.0081}\ \text{PCLME}(-2) \\
&+ \underset{(1.22)}{0.0166}\ \text{PCLME}(-3) + \underset{(2.86)}{0.1105}\ \text{PCLME}(-4)
\end{aligned}
\tag{3.33}
$$

$$\bar{R}^{2} = 0.924 \qquad SEE = 8.21 \qquad DW = 1.42$$

DZ69/70 is the dummy variable for the uncertainties of the then-unconsummated plans for partial nationalization and reorganization of the industry. The mean lag is 3.1, and the short-run price elasticity is 0.11. The average long-run response is much higher, at 1.6.[9]

The second model of investment assumes that real investment (KCMINREAL) is a function of past real net profits (profits after all taxes) and past LME prices. The profits represent the financing aspect, and the prices reflect the expectational aspect. Various lag structures were tried; the best equation for 1964–1976, in terms of expected signs of coefficients and econometric properties, was as follows:

$$\begin{aligned} \text{KCMINREAL} = & -\underset{(-4.15)}{72.5003} + \underset{(6.67)}{0.3240}\ \text{VGPNET65} + \underset{(6.63)}{0.2697}\ \text{VGPNET65}(-1) \\ & + \underset{(6.28)}{0.139}\ \text{PCLME}(-2) - \underset{(-9.27)}{85.3129}\ \text{DZ69/70} \\ & + \underset{(5.82)}{22.9590}\ \text{DZI71/72} \end{aligned} \qquad (3.34)$$

$$R^{-2} = 0.923 \qquad SEE = 4.8777 \qquad DW = 2.66$$

The contemporaneous and the one-period lag real net profits (VGPNET65) were significant, and only the two-period lag LME price entered the equation with a significant positive sign. The one-period lag price was tried, but the coefficient was negative and not statistically significant. The form of the relationship is not surprising, since profits are correlated with prices; thus, profits and prices of the same year do not both enter the equation significantly. The short-run elasticity with respect to net profits is 0.69, and the long-run elasticity is 1.3. The elasticity with respect to the price-lagged two periods is 1.58. The dummy variable DZI71/72 represents the higher-than-normal investment activity in 1971–1972 to compensate for the drop in investment caused by the reorganization of the industry in 1969–1970.

For purposes of simulation, the second formulation of investment determination may be more useful. The first formulation results in investment being exogenous in the model, since the LME prices are taken as given. With the second model, we could examine the effects on investment (and eventually on the long-run performance of the industry and country) of various policies relating to taxation in the industry, output, and employment. The second formulation enables us to get the effects of taxes and different write-off provisions for capital expenditures on the investment decision. This is because the net profits depend on tax rates and the rate of write-off of capital expenditures for tax purposes.

Notes

1. The payments to RST and Anglo-American Corporation for the termination of the contracts were not made until 1977.

2. See *Appendices to Explanatory Statement of Roan Selection Trust Limited,* 30 June 1970, App. H., p. H-1.

3. See F.M. Fisher, P.H. Cootner, and M.N. Baily, "An Econometric Model of the World Copper Industry," *The Bell Journal of Economics and Management Science* 3 (Autumn 1972): 568–609.

4. See L.R. Klein and R. Summers, *The Wharton Index of Capacity Utilization,* Economics Research Unit, University of Pennsylvania, 1966.

5. A more general distributed lag function, the Almon lag, was tried, but it gave results similar to the Koyck lag.

6. There is one major open-pit mine (Chingola) and two smaller ones, which in 1977 provided 54 percent of total output. There are four major underground mines.

7. See B. Hickman, R. Coen, and M. Hurd, "The Hickman-Coen Annual Growth Model: Structural Characteristics and Policy Responses," *International Economic Review* 16 (February 1975):20–37.

8. See Roan Consolidated Mines, Ltd., *Annual Report,* 1970 and 1972. Until 1972, the procedure at the Anglo-American divisions was to finance 67 percent of capital expenditures out of retained profits. See Nchanga Consolidated Copper Mines, Ltd., *Annual Report,* 1973.

9. The very high marginal capital-output ratio, however, results in a lower output elasticity.

Structure and Estimates of the Macro Model

Overview of the Model

The macroeconometric model of Zambia is a disaggregated multisector model. The level of disaggregation is mainly dependent on the availability of data. The model explains the supply and demand of goods, services, and input factors and the process of price formation in the economy. The model is split into seven interrelated blocks: the output, employment, and wages blocks and the demand, foreign, prices, government, and financial-flow blocks. Each block is further disaggregated into subsectors of economic activity. The estimated equations, technical relations, and identities from the micro model of the copper industry, which was discussed in the previous chapter, are used in the related sectors in the macro model. For example, the profits from the copper industry are used to determine mineral revenues in the government sector. Also, the wages paid to employees in the copper industry are important in the determination of wages in the other subsectors of the economy.

The model is largely supply-determined; the gross domestic product is determined as the sum of the economic activities in the various supply subsectors. Demand considerations, however, do play a role in the determination of some sectoral outputs. The model therefore cannot be described as either strictly Keynesian or classical in the sense of being either strictly demands- or supply-dominated; rather, it is more like the disequilibrium models of Clower (1965) and Barro and Grossman (1976).

Another important feature is the disaggregate and highly endogenous government sector. Since the government plays a major role in the economic activities of Zambia, an exogeneous government sector would not produce a very responsive simulation model. An endogenous government sector provides a better linkage between the copper industry and the rest of the economy. The estimated equations presented in this chapter were obtained for the period 1965–1976 unless specified otherwise.

Output, Employment, and Wages

The basis of the macro model is the identity explaining the distribution of value added by end use. We divided the economy into six producing sectors, whose output went to either subsistence or nonsubsistence private consumption (C_s, C_p), government consumption (C_g), total investment (I), net exports (E - M), and changes in inventories (ΔS); that is,

$$GDP = \sum_{i=1}^{6} Y_i = C_s + C_p + C_g + I + (E - M) + \Delta S$$

Zambia's agriculture is of a dual nature, with a subsistence sector producing mostly for own consumption. Savings in this sector are ignored, largely for lack of information, and consumption (C_s) is the same as output. The economy is disaggregated into the following production sectors:

1. Mining and quarrying. Copper mining is the principal activity here. The sector also includes other minerals, such as zinc, lead and cobalt, and coal.
2. Manufacturing. This comprises the activities of enterprises using modern capital equipment to produce a variety of intermediate and final consumption goods.
3. Construction.
4. Transportation and communications. This sector comprises rail and road transport, storage, and posts and telecommunications.
5. Agriculture. Two general types are recognized. Commercial agriculture produces mainly for the market, using comparatively capital-intensive methods. Subsistence or rural agriculture produces mainly for own consumption, employing traditional labor-intensive techniques. In between (not recognized separately in the model) is a group referred to as "emergent" agriculture, which produces mainly for the market and uses both modern capital-intensive techniques and traditional methods.
6. Services sector. This is made up of activities in electricity, water, and gas-supply services; commercial activities such as finance, business services, real estate, and wholesale and retail trade; and community, social, and personal services.

Mining and Quarrying

Output in mining and quarrying is determined as the sum of the value of copper and other mining and quarrying. The output in other mining and quarrying (XMNCOP) is small compared to that of copper; it is made exogenous in the model. From the micro model we obtain the total value of sales and copper (VSXCOP), the value of intermediate inputs (VIMNC), and the total transportation costs of the copper industry (VTTCOST). From these, the value added in the copper industry (VXCOP) is given by

$$VXCOP = VSXCOP - VIMNC - VTTCOST \tag{4.1}$$

This is the same as equation 3.32 in chapter 3. Because of differences between industry accounting and national accounting, VXCOP is not necessarily equal to XGDCOP but is correlated with it. Hence, we made XGDCOP a function of VXCOP and obtained an estimate for the period 1958–1976 as

$$\begin{aligned} XGDCOP = & \underset{(2.96)}{41.533} + \underset{(22.05)}{0.9897}\ VXCOP + \underset{(4.83)}{91.77}\ DZ66/68 \\ & - \underset{(-3.71)}{115.112}\ DZ76 \end{aligned} \quad (4.2)$$

$$\bar{R}^2 = 0.967 \qquad SEE = 30.0 \qquad DW = 1.60$$

DZ66/68 and DZ76 are dummy variables that have a value of 1.0 for the periods 1966–1968 and 1976, respectively, and zero for other periods. The period 1966–1968 was when Zambia was facing severe transportation difficulties. This reduced exports, but since production was not correspondingly curtailed, stockpiles of concentrates were accumulated. In 1976, sales were much higher than output, implying a reduction in stockpiles of concentrates built up in previous years.

Using equation 4.2, then, we get the contribution of mining and quarrying to gross domestic product in nominal kwacha (XMINQ):

$$XMINQ = XGDCOP + XMNCOP \quad (4.3)$$

and the real contribution to GDP (XMINQ65).

$$XMINQ65 = XMINQ / PXMINQ \quad (4.4)$$

where PXMINQ is the deflator for output in mining and quarrying.

Employment in the mining and quarrying sector consists of labor in copper mining and other mining operations. The variables VTLE and VTEE represent local and expatriate labor, respectively, employed in copper mining, and have been determined in the micro model in chapter 3. We take the employment in other mining and quarrying as exogenous to the model. Compared to copper-industry employment, noncopper-mining employment is small. The wages paid in copper mining are assumed to apply to all mining and quarrying. The two categories of wages, of Zambians and of expatriates, have already been determined in the micro model. The expressions for employment in the mining and quarrying sector of Zambians and expatriates, LEMAQZ and LEMAQN, respectively, are

$$LEMAQZ = LEMAQZR + VTLE \quad (4.5)$$

$$LEMAQN = LEMAQNR + VTEE \quad (4.6)$$

where LEMAQZR and LEMAQNR stand for employment of Zambians and expatriates, respectively in noncopper mining and quarrying.

Manufacturing

The manufacturing sector has been the fastest-growing sector in the Zambian economy in terms of contribution to gross domestic product. The sector showed a compound annual growth rate of 7.4 percent between 1965 and 1976. In 1965, the sector contributed just under 7 percent of GDP, and in 1976 its contribution was 11 percent. Its share of total employment was 10.6 percent in 1965 and 11.5 percent in 1976. The Zambian government plays an active role in the manufacturing sector through its ownership and control of major firms.

The manufacturing sector produces consumption goods, such as processed foods, beverages and tobacco, and textiles, and intermediate products, such as fertilizers, chemicals, building materials, and fabricated metal products. A 1970 survey of Zambian industry showed that 48 percent of the 427 manufacturing establishments surveyed produced foods, textiles, and related consumer goods.[1] Of the seventeen firms that employed more than 500 people each, nine produced foods and related products. The government policy is to encourage import-substitution manufacturing industries for essential consumer products, industries that process resource-based raw materials, and industries that produce goods for export. The export of manufactured goods until now has been insignificant.

Zambian manufacturing industries produce goods for which there is already an established and known demand (as import substitutes), and they are protected from foreign competition by tariffs and quantitative restriction on imports. Manufacturing activity is dominated by firms owned by the government through the Industrial Development Corporation (INDECO); the sector cannot be characterized as competitive, nor the firms in it as profit maximizers. INDECO and other parastatals (government-owned firms) are expected to consider other objectives, including self-sufficiency in essential products, import substitution, and provision of employment, in addition to profits in their business decisions.

To determine real value added and employment in the manufacturing sector, we use a production function. Total output y is made a function of the capital stock (k), the labor employed (l), and the input of intermediate goods (m): $y = f(k,l,m)$. We have data for value added (q) as well as for capital stock and labor. Value added can be made a function of k and l only if we assume that the marginal rate substitution between capital and labor is independent of the quantity of intermediate materials.[2] Further, we assume that the production function is of a fixed-coefficient type, with capital as the limiting input. Hence, q becomes a function of k only: $q = f(k)$. The estimate of real value added to manufacturing (XMANF64) is therefore given by

$$\text{XMANF65} = 46.98 + 0.3510\ \text{KSMAN65}(-1)$$
$$(12.4) \quad (11.8)$$

$$R^{-2} = 0.93 \qquad SEE = 6.1 \qquad DW = 1.55 \tag{4.7}$$

The variable KSMAN65 represents the end-of-period capital stocks in manufacturing, expressed in constant 1965 kwacha. Similarly, we estimate an equation for the employment of Zambians in manufacturing (LEMANZ) as a function of manufacturing value added:

$$\text{LEMANZ} = 13.98 + 09.2531\ \text{XMANF65}$$
$$(6.96) \quad (5.31)$$

$$R^{-2} = 0.93 \qquad SEE = 1.6 \qquad DW = 1.2 \tag{4.8}$$

Equation 4.7 implies that the manufacturing sector is always operating at the same level of capacity utilization, which, given the historical performance of the sector, would be near full capacity. This view is reasonable for several reasons. The sector is dominated by government-controlled firms that have social objectives, such as the maintenance of employment. The government is also willing to subsidize production to maintain employment or to keep prices lower than would be dictated by market forces. The sectoral production is usually of essential products with a strong demand. Therefore, insufficient demand is unlikely to be a factor in determination of output. However, production capacity has been idled at various times by shortages of imported inputs and spare parts for machinery and equipment. Transportation difficulties (as already discussed in chapter 1) and occasional scarcity of foreign exchange have been responsible for these shortages.

Political and social considerations play a large role in the determination of wages paid in Zambian industry. As a result of the perceived ill-treatment of African workers by foreign-owned mining and related firms prior to independence, strong and centralized trade unionism exists in Zambia. The mining industry trade union is the strongest, and workers in other industries tend to follow its lead. Thus, the wages paid in other sectors of the economy are related to those paid in mining. We estimate the average annual wage of Zambians in manufacturing (WGAMZ) as a function of the mining wage of Zambians (WGAMQZ) and the lagged consumer price index (PC). The relationship obtained for the period 1960–1976 is

$$\text{WGAMZ} = -414.52 + 0.333\ \text{WGAMQZ} + 11.23\ \text{PC}(-1)$$
$$(-7.9) \quad (5.8) \qquad\qquad (7.7)$$

$$R^{-2} = 0.974 \qquad SEE = 53.58 \qquad DW = 1.46 \tag{4.9}$$

The links between the manufacturing sector and the mining sector are mainly through investment and wages. Government investment, either through direct equity or through loans to parastatal corporations, is a significant part of investment in the noncopper sector, and revenues from copper make much of this investment possible.

Construction

Value added in the construction sector is demand-determined. The nominal value added (XCONST) is obtained as a function of nominal private investment and public investment. The relationship obtained is

$$\begin{aligned} \text{XCONST} = &\ 24.012 + 0.2797\ \text{IGFCP} + 0.0832\ \text{CGDC} \\ &\ (1.90) \quad (5.24) \qquad\qquad (0.54) \end{aligned}$$

$$R^{-2} = 0.83 \qquad SEE = 12.6 \qquad DW = 1.21 \tag{4.10}$$

where IGFCP represents nominal private investment and CGDC is the nominal value of public-sector direct investment. The copper sector affects the construction sector through copper investment, which is part of private investment, and through the mining sector's effect on government revenue, which in turn influences public direct investment. The implication of the estimated relationship is that supply constraints are not operative in the construction sector. This can be justified on the grounds that there is greater opportunity for substitution of labor for capital in construction. The sector's need for labor can be met since it does not need great amounts of skilled labor. The real value added to construction is given by

$$\text{XCONST65} = \text{XCONST/PXCONST} \tag{4.11}$$

where PXCONST is the price deflator for output in construction.

Employment of Zambians in construction (LECZ) is made a function of the real wage in construction and the lagged employment. The relationship estimated for the 1961–1976 period is

$$\begin{aligned} \text{LECZ} = &\ -4.1005 + 2.965\ \text{WGACZ/PC} + 0.6886\ \text{LECZ}(-1) \\ &\ (-0.44) \quad (1.68) \qquad\qquad (4.51) \end{aligned}$$

$$R^{-2} = 0.81 \qquad SEE = 7.48 \qquad DW = 1.31 \tag{4.12}$$

where WGACZ/PC is the real average annual wage rate in construction. Equation 4.12 is a labor-supply equation in the classical framework, although the coefficient of the real wage rate is not significant at the 5 percent level. In an economy with surplus labor, it seems to be contradictory. However, wages in construction are low; only agricultural wages are lower. The work is usually hard and carries a low status, and the employment is, in many cases, not permanent. Thus, the financial incentives have to be high enough to attract even low-paid agricultural workers to construction jobs.

As in the manufacturing sector, the wages in construction are related to those in the mining sector (WGAMQZ) and the lagged consumer price index. For the 1960–1976 period, the fit obtained is

$$\underset{}{\text{WGACZ}} = \underset{(-1.34)}{-59.33} + \underset{(7.41)}{0.3593}\ \text{WGAMQZ} + \underset{(1.65)}{2.029}\ \text{PC}(-1)$$

$$R^{-2} = 0.95 \qquad SEE = 45.1 \qquad DW = 2.4 \qquad (4.13)$$

Transportation and Communications

Output in the transportation and communications sector is determined by the volume of foreign-trade traffic and the level of general economic activity. The foreign trade is measured mainly by the volume of imports, since import volume far exceeds export volume, although export value is generally greater. The relation for nominal value added in transportation (XTRCOM) is given by

$$\text{XTRCOM} = \underset{(0.36)}{3.051} + \underset{(2.54)}{0.0806}\ \text{MGFOB} + \underset{(2.11)}{0.0213}\ \text{XGDMC}$$

$$R^{-2} = 0.82 \qquad SEE = 7.08 \qquad DW = 1.85 \qquad (4.14)$$

where MGFOB is imports of goods (f.o.b.) and XGDMC is the nominal gross domestic product.

Zambia has suffered from a lack of adequate transportation routes for its foreign trade. It depends on neighboring countries for access to the sea. Political problems in these countries have severely limited Zambia's capacity in international trade since 1966. The mining sector influences transportation through its effect on the ability to import, on the volume of mining-related imports, and on the overall level of economic activity. The real value added in transportation and communications is given by

$$\text{XTRCOM65} = \text{XTRCOM/PXTRCOM} \qquad (4.15)$$

where PXTRCOM is the deflator for output in transportation and communications.

The equation for employment in the transportation and communications sector is given by

$$\begin{aligned} \text{LETACZ} = {} & \underset{(0.10)}{0.5033} + \underset{(1.18)}{0.1511}\ \text{XTRCOM 65} + \underset{(5.74)}{0.6504}\ \text{LETACZ}(-1) \\ & + \underset{(2.46)}{4.2142}\ \text{DZ66} \end{aligned} \tag{4.16}$$

$$\bar{R}^2 = 0.81 \qquad SEE = 1.26 \qquad DW = 1.66$$

where DZ66 is a dummy variable representing disruptions in the transportation sector in 1966 caused by Rhodesia's unilateral declaration of independence (UDI).

The average annual wage of Zambians in transportation and communications (WGATCZ) is a function of the wage in mining and the lagged consumer price index. The relationship obtained for 1960–1976 is

$$\text{WGATZ} = \underset{(-6.23)}{-748.76} + \underset{(2.89)}{0.3796}\ \text{WGAMQZ} + \underset{(5.91)}{19.64}\ \text{PC}(-1) \tag{4.17}$$

$$\bar{R}^2 = 0.94 \qquad SEE = 122.2 \qquad DW = 2.3$$

Services

Output in the services sector is a function of real government consumption expenditures (BCGC65) and the urban population (POPU). The services sector is dominated by the central-government activity in urban areas. The estimated equation for the real value added by the services sector (XSER65) is given by

$$\text{XSER65} = \underset{(1.87)}{27.68} + \underset{(2.73)}{0.6754}\ \text{BCGC65} + \underset{(3.45)}{114.989}\ \text{POPU} \tag{4.18}$$

$$\bar{R}^2 = 0.97 \qquad SEE = 12.1 \qquad DW = 2.6$$

Employment of Zambians in services is a function of real value added and is given by

$$\text{LESERZ} = \underset{(0.11)}{0.7371} + \underset{(16.08)}{0.3652}\ \text{XSER65}$$

$$\bar{R}^2 = 0.96 \qquad SEE = 5.3 \qquad DW = 3.0 \tag{4.19}$$

The wages of Zambians in the service sector (WGASZ) is estimated as a function of the wages of Zambians in mining (WGAMQZ) and the lagged consumer price index (PC). For the 1960–1976 period, it is given by

$$\underset{}{WGASZ} = \underset{(-7.45)}{-622.76} + \underset{(4.72)}{0.4319}\ WGAMQZ + \underset{(5.63)}{13.02}\ PC(-1) \qquad (4.20)$$

$$\bar{R}^2 = 0.956 \qquad SEE = 85.1 \qquad DW = 1.34$$

Agriculture

The agricultural sector comprises two subsectors, commercial agriculture and subsistence agriculture. The principal commodity produced is maize. Other products include tobacco, groundnuts (peanuts), sugar cane, millet, poultry, dairy, and meat. Most agricultural output is to meet domestic consumption, although maize and tobacco have been exported in the past. The export of maize has fallen off because of increased domestic needs, and tobacco, which is grown mainly for the export market, has shown a sluggish performance. In 1965, the subsistence sector contributed 81 percent of value added in agriculture, but by 1976 the contribution had fallen to 61 percent.

The commercial agriculture sector, consisting of large state and private farm estates, employs capital-intensive techniques and sells the output through official marketing boards, which also determine the price levels. The real value added in commercial agriculture (XAFFCS65) is determined as a function of the capital stock in agriculture and the price of output. The producer price of maize (PPM), which is the dominant commodity, is used as a proxy for the unit value of total output. The estimated relationship is

$$XAFFCS65 = \underset{(4.06)}{14.048} + \underset{(2.43)}{0.1220}\ KSCA65(-1) + \underset{(2.42)}{2.3744}\ PPM + \underset{(3.77)}{9.4475}\ DZ72/76 \qquad (4.21)$$

$$\bar{R}^2 = 0.95 \qquad SEE = 2.43 \qquad DW = 1.61$$

where KSCA65 is the end-of-period capital stock employed in agriculture. DZ72/76 is a dummy variable for 1972–1976. This represents the normalization of production in many expatriate-owned commercial estates. Following independence and the UDI crisis in Rhodesia, the number and volume of operations of expatriate-owned farms declined, but they seemed to stabilize from 1971 on. The elasticities of output with respect to capital and the price of maize are 0.24 and 0.28, respectively.

Output in subsistence agriculture is mainly for own consumption. A census of agriculture shows that, for 1970–1971, noncommercial farmers sold or bartered 36 percent of their output of maize.[3] However, the noncommercial farmers account for a large share of total output. The prices at which they sell the marketed portion of their output are based on the producer prices received by the commercial farmers. The output in subsistence or rural agriculture (XAFFS65) is therefore made a function of the producer price of maize (PPM) and the rural population (POPR); the latter is a proxy for the labor force in rural agriculture. The estimated relationship is

$$\begin{aligned} \text{XAFFS65} &= \underset{(5.19)}{52.75} + \underset{(1.34)}{0.6240}\ \text{PPM} + \underset{(2.27)}{9.1804}\ \text{POPR} \\ \bar{R}^2 &= 0.934 \qquad SEE = 0.532 \qquad DW = 1.40 \end{aligned} \tag{4.22}$$

The elasticity of output with respect to the producer price of maize is 0.033; with respect to rural population, it is 0.33. The price elasticity with respect to PPM is much lower than that of the commercial sector (0.24). This is not surprising, since the commercial sector is market-oriented and much better organized to respond to changes in market conditions. The elasticity with respect to rural population is low. This may be because migration to urban centers is depriving the rural areas of the adult young (especially men), the educated, and the more physically, mentally, or materially endowed. The rural population then gets increasingly dominated by the very young, the old, and women.

To determine employment in commercial agriculture (LEAGZ), we estimate the labor-output ratio (LEAGZ/XAFFCS65). This is made a function of the capital-output ratio and a time trend (TIME), the latter to account for exogenous changes in productivity. The equation obtained is

$$\begin{aligned} \text{LEAGZ/XAFFCS65} &= \underset{(5.04)}{1.7671} + \underset{(0.43)}{0.0562}\ \text{KSCA65(–1)/XAFFCS65} \\ &\quad - \underset{(-1.30)}{0.0363}\ \text{TIME} - \underset{(-2.24)}{0.3502}\ \text{DZ72/76} \\ \bar{R}^2 &= 0.878 \qquad SEE = 0.11 \qquad DW = 1.45 \end{aligned} \tag{4.23}$$

The relationship shows that capital is a complement to labor, holding productivity constant. When the equation was estimated without the time trend, the coefficient of the capital-output ratio was negative and mildly significant (*t* value = 1.91), implying that capital and labor are substitutes. We chose the given relationship because it has better overall statistical and simulation properties. The growth in productivity with time may be a result of economies of scale and increased use of capital equipment. The dummy variable DZ72/76 represents the same circumstances as in equation 4.21, the effect here being an increase in productivity.

The wage rate of Zambians in commercial agriculture is related to wages paid in the mining industry (WGAMQZ) and a measure of the productivity of labor given by the output-labor ratio (XAFFCS65/LEAGZ).

$$\underset{}{\text{WGAAFFZ}} = \underset{(-0.77)}{-55.51} + \underset{(3.66)}{0.2342}\ \text{WGAMQZ} + \underset{(1.4)}{66.1426}\ \text{XAFFCS65/LEAGZ} \qquad (4.24)$$

$$R^{-2} = 0.743 \qquad SEE = 46.8 \qquad DW = 1.84$$

where WGAAFFZ is the average annual wage of Zambians in agriculture. Wages rise as productivity increases. The cost-of-living adjustment is not as significant in agriculture as in other sectors because of the absence of strong labor-union activities.

Employment and Wages of Expatriates in Nonmining Sectors

The total employment of expatriates (LEEXPR) in all nonmining sectors is made a function of the compensation of expatriates in pounds sterling (WGAEXPR * EXCH), where WGAEXPR is the expatriate average annual compensation in kwacha and EXCH is the exchange rate in pounds sterling per kwacha. The compensation of expatriates is related to the average annual compensation of expatriates in mining (WGAMQN) and the lagged value of the dependent variable. The estimated relations of LEEXPR and WGAEXPR for the period 1960–1976 are

$$\text{LEEXPR} = \underset{(53.88)}{26.2115} - \underset{(-6.97)}{0.0014}\ \text{WGAEXPR} * \text{EXCH} \qquad (4.25)$$

$$R^{-2} = 0.75 \qquad SEE = 0.718 \qquad DW = 2.03$$

$$\text{WGAEXPR} = \underset{(-0.74)}{-261.633} + \underset{(2.43)}{0.1466}\ \text{WGAMQZ} + \underset{(13.39)}{0.8570}\ \text{WGAEXPR}(-1) \qquad (4.26)$$

$$R^{-2} = 0.95 \qquad SEE = 232.9 \qquad DW = 1.77$$

In chapter 3, the employment of expatriates in mining was estimated. As in equation 4.25 it was negatively related to the wage rate, indicating a demand function rather than the expected supply function. It was pointed out that the reported wage rate does not accurately measure the compensation of expatriates.

In addition to their wages, expatriates also receive both monetary and nonmonetary allowances, such as lump-sum payments at the expiration of their contracts and free housing. Employment of expatriates is inelastic with respect to the wage rate, the elasticity being 0.14. Increases or decreases in the basic wage seems to have little effect on the number of expatriates employed.

Gross Domestic Product and Total Employment

From the sectoral value added, we sum to get the gross domestic output of economy:

$$\text{XGDMC} = \text{XMINQ} + \text{XMANF} + \text{XCONST} + \text{XTRCOM} + \text{XAFFS} + \text{XAFFCS} + \text{XSER} \qquad (4.27)$$

$$\text{XGDMC65} = \text{XMINQ65} + \text{XMANF65} + \text{XCONST65} + \text{XTRCOM65} + \text{XAFFS65} + \text{XAFFCS65} + \text{XSER65} \qquad (4.28)$$

where XGDMC and XGDMC65 represent gross domestic product at market prices and constant 1965 kwacha, respectively. Similarly, we obtain the total employment of Zambians (LEINDZ) and expatriates (LEINDN) by summing the sectoral employment:

$$\text{LEINDZ} = \text{LEMAQZ} + \text{LEMANZ} + \text{LECZ} + \text{LETACZ} + \text{LESERZ} + \text{LEAGZ} \qquad (4.29)$$

$$\text{LEINDN} = \text{LEMAQN} + \text{LEEXPR} \qquad (4.30)$$

Employees' Nonwage Benefits

In addition to wages, employees receive other benefits, such as employer's contribution to pension plans, housing and transportation subsidies or allowances, and bonuses. These benefits differ for Zambians and expatriate employees. We make the total real nonwage benefits (YWO/PC) a function of total employment of Zambians (LEINDZ) and expatriates (LEINDN). The estimated relation for the period 1960–1976 is

$$\underset{}{\text{YWO/PC}} = \underset{(-4.05)}{-4.4245} + \underset{(8.01)}{0.0095}\ \text{LEINDZ} + \underset{(3.05)}{0.0935}\ \text{LEINDN} + 1.4602\ \text{DZ76} \qquad (4.31)$$

$$\bar{R}^2 = 0.88 \qquad SEE = 0.28 \qquad DW = 1.58$$

The Aggregate-Demand Block

This section explains the aggregate-demand side of the economy. The components to be explained are private-consumption expenditures, public or government-consumption expenditures, investment in the nonmining private sector, government direct-investment expenditures, and change in inventories. Investment in copper mining was explained in the micro model.

Private Consumption

In economic theory, there are several competing theories of consumption, including those of Keynes, Modigliani-Ando-Brumberg, and Milton Friedman.[4] The essential postulate of these theories is that consumption is a function of some form of aggregate income. Other modifications include the consideration of the roles of peak incomes, total assets, money balances, and distribution of income by wage and nonwage components. In the Keynesian model, aggregate consumption is a function of aggregate income, with both variables deflated in terms of wage units. Modigliani-Ando-Brumberg's life-cycle hypothesis posits that a consumer adjusts its consumption to the average income expected over its life time. Friedman divides income into its permanent and transitory components and makes consumption a function of permanent income. To make the model operational for time-series data, Friedman assumes that aggregate permanent income is a function of a weighted average of present and past disposable income. In the Zambian economy, where aggregate income is subject to variations caused by fluctuations in copper prices, a permanent income model of consumption seems appropriate. In the short run, however, Zambian disposable incomes are shielded from the fluctuations of copper prices because government revenues and copper-company profits absorb most of the resulting variations in income, and dividend payments by the companies accrue either to the government or to foreign shareholders.

Private-consumption expenditure is made up of urban and rural consumption. There is neither sufficient nor up-to-date data on rural consumption to allow for any estimation of that component. We therefore make the simplifying assumption that rural consumption is equal to the output in subsistence agriculture. This assumption is not unrealistic, since a comparison of the few years in which there are data for the two series shows that their values are reasonably close. The model for urban-consumption expenditures makes real per capita consumption a function of real disposable income and lagged real per capita consumption. In order to take into account the effect of income distribution, real disposable income is divided into its wage and nonwage components. The inclusion of the lagged dependent variable is a permanent income-model feature that captures the effect of past disposable income.

As a result of the high share of wage income earned by expatriates in Zambia, the effect of income transfer abroad out of wage income is taken into account.

Disposable wage income is therefore defined as total wage income less taxes and net income transferred abroad. Per capita disposable income is therefore given by

$$YWDRPC = (YW - CGIT - YNOEX)/POPU/PBCPC \qquad (4.32)$$

where YW is wage income, CGIT is personal income taxes, YNOEX is net transfers, POPU is urban population, and PBCPC is the deflator for private-consumption expenditures. Similarly, the effect of transfers abroad from nonwage income (profit) is taken into account in determining disposable nonwage income, since these transfers are large in magnitude. We therefore have

$$YPROFDRPC = (YPROF - XAFFS - YNIEX)/POPU/PBCPC \qquad (4.33)$$

where YPROFDRPC is the per capita real disposable nonwage income, YPROF is the operating surplus, XAFFS is the output in subsistence agriculture, and YNIEX is the net investment income transferred abroad.

Using these definitions of disposable incomes, the following relationship was obtained for per capita urban real-consumption expenditure (UBCP) for the period 1966–1976

$$\begin{aligned} UBCP = &- \underset{(-1.53)}{35.5999} + \underset{(5.21)}{0.6872}\ YWDRPC + \underset{(2.70)}{0.22085}\ YPROFDRPC \\ &+ \underset{(3.02)}{0.3644}\ UBCP(-1) \qquad (4.34) \end{aligned}$$

$$\bar{R}^2 = 0.966 \qquad SEE = 11.5 \qquad DW = 2.14$$

The short-run marginal propensity to consume (MPC) out of wage income is 0.69, and the short-run MPC out of nonwage income is 0.22. The long-run MPCs out of wage and nonwage incomes are 1.08 and 0.35, respectively. The weighted average (using the mean values as weights) short-run and long-run MCPs out of total per capita disposable income are 0.59 and 0.78. The weighted average propensity to consume out of total disposable income is 0.79.

The MPC out of nonwage income seems low. There are possible explanations for this. Because of a lack of data, it was not possible to subtract company taxes on profits from the operating surplus to get disposable nonwage income. The measurement of operating surplus and therefore disposable nonwage income may be subject to greater measurement errors than that of wage income. This will tend to bias the coefficient of disposable nonwage income downward. Most nonwage income accrues to government or foreign-owned companies. Profits of foreign-owned companies that are not sent abroad as dividends are retained for reinvestment or working capital. The profits of government-owned companies are not distributed for private consumption. In the absence of access to a developed capital market, Zambian entrepreneurs would have to rely on their own savings for investment funds. The nonexistence of a developed capital market and credit system may also help to explain the low short-run MPC out of wage income. Since consumers have to pay cash for most of their purchases, they have to save up for most major purchases and thus may not be able to make their desired pruchases when their income rises. In the long run, purchases catch up with income, so that the MPC gets near unity. Another possible explanation for the low short-run MPC out of wage income is the uncertainty attached to future income because of the fluctuations in copper prices.

Total real urban-consumption expenditures (UBC65) and total real private-consumption expenditures (BCPC65) are obtained as

$$\text{UBC65} = \text{UBCP} * \text{POPU} \tag{4.35}$$

$$\text{BCPC65} = \text{UBC65} + \text{XAFFS65} \tag{4.36}$$

where XAFFS65 is the real output in subsistence agriculture.

Government-Consumption Expenditures

Government-consumption expenditures (BCGC) are a function of budgeted expenditures on general services (CGGS) and social services (CGSS). These are the two major components of government expenditures. The estimated relationship is

$$\begin{aligned} \text{BCGC} = \underset{(-0.42)}{-8.119} + \underset{(3.62)}{2.6049}\ \text{CGSS} + \underset{(1.68)}{0.6178}\ \text{CGGS} \end{aligned}$$

$$R^{-2} = 0.966 \qquad SEE = 21.1 \qquad DW = 1.15 \tag{4.37}$$

The real government-consumption expenditures (BCGC65) are obtained by deflating with the price deflator for government-consumption expenditures (PBCGC):

$$\text{BCGC65} = \text{BCGC/PBCGC} \tag{4.38}$$

Government Direct Investment

Government direct-investment expenditures depend on lagged government recurrent revenues and the change in government recurrent revenues.

$$CGDC = \underset{(1.46)}{18.4539} + \underset{(1.60)}{0.70}\,(CGRR - CGRR(-1)) + \underset{(6.44)}{0.2141}\,CGRR\,(-1) + \underset{(3.30)}{47.8078}\,DZ68 \qquad (4.39)$$

$$\bar{R}^2 = 0.82 \qquad SEE = 13.74 \qquad DW = 1.23$$

where CGDC is the nominal value of government direct-investment expenditures and CGRR is current government revenues. DZ68 is a dummy variable representing increased expenditures to take up the shortfall in private investment following the announcement of the 51 percent government takeover of major private nonmining firms and the extraordinary expenditures caused by the construction of the Tanzam railway.

Government Loans and Investments

Zambia has a large parastatal sector, which is classified as part of the private sector in statistical analyses. Some of the companies are wholly owned by the state, but in some the state is in partnership with private entrepreneurs. The government provides investment finance to the parastatal sector in the form of either loans or equity (investments). The government supply of loans and investments (CGLIC) is a function of lagged government saving represented by the recurrent budget surplus (CGBS) and a time trend (TIME).

$$CGLIC = -\underset{(-3.7)}{77.49} + \underset{(7.29)}{0.1854}\,CGBS(-1) + \underset{(6.25)}{5.7821}\,TIME + \underset{(2.92)}{24.7924}\,DZ68 - \underset{(-3.06)}{25.0189}\,DZ71 \qquad (4.40)$$

$$\bar{R}^2 = 0.884 \qquad SEE = 7.26 \qquad DW = 2.08$$

The dummy variable DZ68 represents the same effect as in equation 4.39. The dummy variable DZ71 represents the effect of large financial commitment by the government in the takeover of the mining companies in 1970, which probably affected its ability to provide financing to parastatals in 1971.

These three categories of government expenditure—consumption, direct investment, and loans and investments—are effected by copper prices and copper-industry activity through the effect of copper on government revenues. Government-consumption expenditure is indirectly dependent on government revenues, since the variables that determine it are dependent on government revenues. Loans and investments depend on government saving, which is largely determined by recurrent revenues. Loans and investments are not a component of final demand but are explained here because they play a crucial role in the determination of private-investment demand.

Private Noncopper Investment

The most widely used model of investment in econometric work is based on the neoclassical theory of investment developed by Dale Jorgenson.[5] The model is developed from the theory that assumes that the objective of a firm is to maximize its present value. Jorgenson obtains his result by maximizing the present value of net receipts of the firm subject to the firm's strictly convex production function and the identity that makes net investment equal to gross investment minus replacement investment. He uses the marginal conditions for the demand for labor services and capital services to obtain a complete neoclassical model of optimal capital accumulation, which includes a shadow price for capital services. The shadow price, also called the user cost of capital, is a function of the price of investment goods, the interest rate, the depreciation rate, and the rate of change of the price of investment goods.

The objective in modeling investment through the neoclassical model is to relate investment to the optimal or desired capital stock. This can be accomplished by relating changes in actual capital stock (net investment) to a distributed-lag function of changes in desired capital stocks. This procedure results in the flexible-accelerator mechanism. Alternative models of investment behavior use different determinants of the desired capital stock. Variables that have been used to explain desired capital stock include capacity utilization, internal funds, cost of external funds, and output or sales.[6] Jorgenson's neoclassical model makes desired capital stock a function of output and the user cost of capital. Therefore, investment is a function of the cost of capital, the rate of capital depreciation, the cost of capital goods and the rate of change of the cost of capital goods, and the taxation, all through the user cost of capital. Use of models of investment behavior based on the neoclassical model and the flexible-accelerator principle has provided good econometric estimates for the United States. However, results for other countries have not been nearly so good; in some cases they have been very poor.

The neoclassical model of investment may not be appropriate for Zambia because the underlying theory is based on assumptions that may not be tenable in that situation. First, there is the assumption that firms' investment decisions are based on the maximization fo present value. In Zambia, the government exerts a great influence in investment decisions of firms through its part-ownership of major means of production and its control of financial resources. The government is committed to broad social and political objectives, which influences the decisions of the government and the firms it owns. It is also unlikely that the foreign-owned firms are maximizing present value; they are probably more interested in minimizing the payback periods becuase of the perceived high risk of doing business. Second, the use of the marginal conditions implies that the firm will keep investing or divesting until those marginal conditions are met. This assumes that there is a functioning and efficient capital market and a capital-goods market. The firm can borrow as much money as it needs in excess of its own resources without affecting the market rate of interest. It can also buy or sell as much capital goods as it desires without affecting the price of capital goods. In the absence of a capital market and a capital-goods market, as is the case for Zambia, the assumed marginal conditions are not likely to hold.

In determining a model of private investment in Zambia, we must take into consideration the special characteristics of the economy. The role of the government as a major saver and dispenser of investment funds to the private sector must be considered. We should also consider such factors as output, profits, availability of credit, foreign exchange to purchase investment goods, and the lack of a developed capital market. Interest rates will not be a factor, since they are controlled by the government, and the principal criterion for their levels seem to be to keep the costs of government borrowing at manageable levels. Historically, Zambian interest rates have been low, and, given the high inflation rates in the 1970s, this implies that, for much of the period, real rates were negative. Interest rates are therefore not a deterrent to borrowing. However, the government restricts private borrowing by reducing commercial banks' excess liquidity.

Our model of real private investments considers the following variables: real output, change in real output, real government loans to and investments in the private sector, real money supply, lagged profits, foreign-exchange reserves, and lagged capital stocks. Various combinations of these variables were tried. Profits, capital stocks, foreign-exchange reserves, and change in real output were consistently insignificant. Change in real output, foreign-exchange reserves, and profits did not even have the expected positive sign. Government loans and investments were consistently significant. Real output and real money supply were not significant when used in the same equation but were significant when used separately. This is probably the problem of multicollinearity. We therefore have two estimates for the real investment by the noncopper private sector (IGFCPNM65): one makes IGFCPNM65 a function of government loans and investments (CGLIC65) and real output (XGDMC65), and the other makes it a function of CGLIC65 and real money supply (MSQM/PGDPNM65). The estimated equations are

$$\begin{aligned} \text{IGFCPNM65} = {} & \underset{(.39)}{14.292} + \underset{(4.89)}{1.393}\ \text{CGLIC65} + \underset{(1.9)}{00896}\ \text{XGDM65} \\ & - \underset{(-4.5)}{61.593}\ \text{DZ68} - \underset{(-2.7)}{26.478}\ \text{DZ73/74} \end{aligned} \tag{4.41}$$

$$\bar{R}^2 = 0.832 \quad SEE = 10.9 \quad DW = 3.30$$

$$\begin{aligned} IGFCPNM65 = {} & \underset{(5.48)}{64.723} + \underset{(2.87)}{1.0298}\ CGLIC65 + \underset{(2.25)}{0.1772}\ MSQM/PGDPNM \\ & - \underset{(-3.62)}{52.5715}\ DZ68 - \underset{(-3.12)}{32.684}\ DZ73/74 \end{aligned} \qquad (4.42)$$

$$R^{-2} = 0.85 \qquad SEE = 10.3 \qquad DW = 3.50$$

The dummy variable DZ68 represents the effect of the announcement of government takeover of a 51 percent share of major private enterprises in 1968. This had the effect of dampening private investment. The dummy variable DZ73/74 represents the effect of the closures of the border between Zambia and Rhodesia in 1973. This impeded the flow of imports, including capital goods, into Zambia until alternative routes were found. In equation 4.42, real money supply is a proxy for credit availability. In the absence of organized markets for equity capital, with commercial banks as the dominant financial institutions, commercial-bank credits will be an important determinant of private investment, and even more so when the government owns and controls a large part of the private sector. These parastatal firms are probably not very profitable and as a result need outside financing of their investments. Their links to the government enhance their credit worthiness even when their profit record might not. Commercial-bank credit depends on money supply, and investment therefore depends on money supply.

Changes in Inventories

To complete the determination of aggregate demand, we provide a simple model of changes in inventories. It is made a function of copper inventories, which are defined as the difference between finished production and export and consist mainly of work-in-process inventories and pipeline stocks within Zambia. Another explanatory variable used is the change in final consumption (DFINCI). The estimated equation is

$$\begin{aligned} INCHS65 = {} & \underset{(0.73)}{5.893} + \underset{(0.76)}{0.3696}\ (VTOTP - EXCOPQT) + \underset{(2.65)}{0.2338}\ DFINCI \\ & + \underset{(3.65)}{103.2973}\ DZHALF73/74 \end{aligned} \qquad (4.43)$$

$$R^{-2} = 0.725 \qquad SEE = 23.09 \qquad DW = 2.23$$

where INCHS65 is change in real inventories, VTOTP is the finished copper output less exports (EXCOPQT), DZHALF73/74 is a dummy variable having values of 0.5 and 1.0 in 1973 and 1974, respectively, and zero in other years. It represents the effect of the Rhodesian border closure, which has already been discussed.

Sectoral Investment and Capital Stocks

There are six sectors of economic activity in the model, as outlined at the beginning of this chapter. For each sector we need to derive values of investment and capital stocks. The data that are available on investment by sector are not sufficient to allow for the estimation of sectoral investment. We therefore assign each sector a share of gross real investment. Since the copper-sector investment is determined endogenously in chapter 3, the gross investment less investment in mining is shared out among the other sectors. Therefore:

$$I_i = S_i (I - I_1) \qquad 0 < S_i < 1.0$$

where I_i is the investment in sector $i (i = 2, 3, \ldots, 6)$, I_1 is the investment in copper mining, I is the gross real investment, and S_i is the share of sector i in gross noncopper investment.

Since

$$\sum_{i=2}^{6} S_i = 1,$$

we specify the shares of four sectors: manufacturing, construction, transportation and communications, and commercial agriculture. The share of services (S_6) is obtained as a residual.

Having obtained sectoral investment, the capital stocks for each sector can be derived. We assume a constant rate of depreciation of capital, and, given initial capital stock and gross investment for each period, we obtain sectoral capital stocks as follows:

$$K_t^i = (1 - d_i) K_{t-1}^i + I_i$$

where K_t^i is the real capital stock for sector i in period t, d_i is the depreciation rate of capital in sector i in period t, and I_i is the real investment in sector i in period t.

Foreign Trade

Exports

The major export commodity of Zambia is copper, usually accounting for over 90 percent of total exports. Other mineral exports are lead, zinc, and cobalt, and agricultural exports include tobacco and maize. The exports of copper are determined endogenously in the model, while the other exports are exogenous. In the micro model (chapter 3) the sales of copper (VTESP) were determined. Since almost all copper produced is exported, sales and exports should be close.

They are slightly different because sales and export transactions are recorded at different times. To reconcile them, we make the exports of copper in metric tons (EXCOPQT) a function of sales of copper and obtain the following relationship for the period 1954–1976:

$$\underset{}{\text{EXCOPQT}} = \underset{(.48)}{4.3172} + \underset{(65.3)}{0.9922}\ \text{VTESP} + \underset{(-10.3)}{95.3142}\ \text{DZ69} \tag{4.44}$$

$$R^{-2} = 0.995 \qquad SEE = 8.3 \qquad DW = 1.62$$

The export price of copper is related to the London Metal Exchange price; we therefore relate the export price in kwacha per metric ton (EXCPRICE) to the LME price in kwacha (VZPRICE) and get, for the 1957–1976 period,

$$\text{EXCPRICE} = \underset{(-2.07)}{-46.5841} + \underset{(34.9)}{0.994}\ \text{VZPRICE} - \underset{(-12.6)}{268.36}\ \text{DZ64/66} \tag{4.45}$$

$$R^{-2} = 0.985 \qquad SEE = 33.4 \qquad DW = 2.70$$

where DZ64/66 is a dummy variable for the period 1964–1966, when the exports were based on producer prices rather than on the LME price.

Imports

Zambia imports consumer goods, capital goods, and services. In the model, imports of goods are dissaggregated by Standard Internation Trade Classification (SITC) categories. M01G69 is the real imports in SITC categories 0 and 1, which comprise beverages, food, and tobacco. M01G69 is made a function of real final demand (FINCI), the real output of agriculture, and the price index of imports in the category (PM01G).

$$\begin{aligned} \text{MOIG69} = {} & \underset{(2.24)}{27.82} + \underset{(2.39)}{0.0447}\ \text{FINCI} - \underset{(-1.64)}{0.2714}\ \text{XAFF65} \\ & - \underset{(-1.17)}{4.8006}\ \text{PM01G} + \underset{(5.02)}{16.788}\ \text{DZ71/72} \end{aligned} \tag{4.46}$$

$$R^{-2} = 0.811 \qquad SEE = 3.15 \qquad DW = 1.71$$

The dummy variable DZ71/72 is for the drought in Zambia that affected agricultural production. The negative sign for real output in agriculture (XAFF65) implies that increased output in domestic agriculture diminishes the need for imports.

The second category of imports (M24G69) corresponds to SITC categories 2 and 4, which are made up of raw and crude materials, excluding fuels. Since they are imported mainly for further processing into finished or semifinished goods, M24G69 is made a function of real output in manufacturing (XMANF65) and a measure of the country's ability to import (MCAPRES). For the period 1966–1975, this yields

$$\underset{}{M24G69} = \underset{(2.94)}{3.2659} + \underset{(3.42)}{0.0343}\ XMANF65 + \underset{(5.22)}{0.0145}\ MCAPRES \tag{4.47}$$

$$\bar{R}^2 = 0.752 \qquad SEE = 0.55 \qquad DW = 1.9$$

MCAPRES is defined as NINTR(−1)/PMGS, where NINTR is net international reserves and PMGS is the price index of imports of goods and services.

A third category (M3G69) is the import of electricity and mineral fuels. The explanatory variables are the corresponding price index and the ability to import. The equation obtained for the 1966–1975 period is

$$M3G69 = \underset{(6.01)}{31.716} - \underset{(-1.81)}{3.1199}\ PM3G + \underset{(0.55)}{0.0128}\ MCAPRES - \underset{(-1.51)}{6.6779}\ DZ72 \tag{4.48}$$

$$\bar{R}^2 = 0.37 \qquad SEE = 4.15 \qquad DW = 1.15$$

An attempt to link the imports of fuel to an economic-activity variable did not yield any significant results.

The fourth category (MCH69) corresponds to SITC category 5, which consists of chemicals. The major component of this section is fertilizer. This category is explained by real output in agriculture (XAFF65) and the ability to import and for the 1966–1976 period yields

$$MCH69 = -\underset{(-3.60)}{42.9313} + \underset{(6.26)}{0.5777}\ XAFF65 + \underset{(0.72)}{0.0099}\ MCAPRS \tag{4.49}$$

$$\bar{R}^2 = 0.819 \qquad SEE = 2.74 \qquad DW = 1.97$$

This formulation may suffer from a simultaneous-equation problem when estimated with ordinary least squares. It can be argued that output in agriculture is dependent on the imports of fertilizers, not the reverse, although one could also argue with equal justification that planned imports of fertilizers depend on planned output in agriculture. It is not clear which is the dominating direction of causality in the Zambian context.

The import of machinery and transportation equipment (MMATE69) corresponds to SITC category 7. It is estimated as a function of real gross fixed-capital formation and the corresponding import price index. The equation obtained for the period 1964–1975 is

$$\begin{array}{llll} \text{MMATE69} = 37.719 & + 0.4559\ \text{IGFC65} & - 24.0898\ \text{PMMATE} \\ \qquad\qquad\ \ (3.51) & \quad (8.63) & \quad (-2.64) \\ \quad - 21.3714\ \text{DZ73} & & \\ \quad\ (-2.43) & & \end{array} \tag{4.50}$$

$$R^{-2} = 0.883 \qquad SEE = 8.10 \qquad DW = 2.90$$

where IGFC65 is the real gross fixed-capital formation and PMMATE is the price index of imports of machinery and transportation equipment. DZ73 is a dummy variable for 1973, presumably linked to the closure of the border with Rhodesia, which was the major route for imports. The ability to import (MCAPRES) was tried as an independent variable, but it was not significant and did not have an expected positive sign. This is not surprising because, in Zambia, imports of investment goods are given priority in the allocation of foreign exchange in periods when foreign exchange is rationed.

The two remaining categories of imports of goods—manufactured goods (MMAG69) and miscellaneous manufactures (MMMA69)—both correspond to SITC categories 6 and 8. The estimated equations for them for the period 1964–1976 are

$$\begin{array}{lllll} \text{MMAG69} = & 11.48 & + 0.0856\ \text{FINCI} & - 11.3086\ \text{PPMAG} & - 13.7277\ \text{DZ73} \\ & (1.54) & \quad (7.23) & \quad (-2.83) & \quad (-2.98) \end{array} \tag{4.51}$$

$$R^{-2} = 0.844 \qquad SEE = 4.33 \qquad DW = 2.13$$

$$\begin{array}{lllll} \text{MMMA69} = & 31.4701 & + 0.0124\ \text{FINCI} & - 13.6786\ \text{PMMA} & - 4.3969\ \text{DZ73} \\ & (7.97) & \quad (1.82) & \quad (-4.33) & \quad (-1.79) \end{array} \tag{4.52}$$

$$R^{-2} = 0.699 \qquad SEE = 2.30 \qquad DW = 2.84$$

where FINCI is real final demand, PMMAG is the price index of imports of manufactured goods, and PMMA is the price index of imports of miscellaneous manufactures.

Finally, real import of services (MSS65) depends on the volume of merchandise imports and a time trend.

$$\begin{array}{lllll} \text{MSS65} = & - 49.0923 & + 0.2996\ \text{MG65} & + 4.0378\ \text{TIME} & + 30.6081\ \text{DZ68} \\ & (-1.02) & \quad (1.74) & \quad (2.57) & \quad (1.77) \end{array} \tag{4.53}$$

$$R^{-2} = 0.599 \qquad SEE = 15.6 \qquad DW = 1.73$$

where MG65 is the real import of merchandise and TIME represents a time trend.

Price Deflators

The price deflators to be determined are those of imports and exports and those for the components of aggregate demand and supply. In addition, two consumer price indexes, for expatriates and for Africans, will be determined.

Aggregate-Supply Deflators

A price deflator is determined for each of the six sectors of economic activity. Separate price deflators are used for output in commercial and subsistence agriculture. The procedure adopted is to determine an aggregate-supply price level and to derive the deflators of the components from it. However, differences in price determination between the mining sector and the nonmining sector have to be taken into account. Since almost all mining output is exported, it does not substantially contribute to domestic demand. Activity in mining does influence the demand for output of the other sectors. We derive the deflators for output in nonmining sectors (except agriculture) as a function of the price index of aggregate output in all nonmining sectors.

In advanced industrial economies, there are two alternative theories of output-price determination. One holds that producers price their output on the basis of a markup on unit labor or total costs, with adjustments for pressures on capacity and input costs. The other theory holds that monetary expansion is the cause of inflation. In developing economies, cost factors and excess-demand conditions may both be important for price-level determination. Excess demand is usually caused by expansionary governmental monetary and fiscal policies. Expansionary fiscal policy is usually financed by the creation of new money. Excess demand occurs from expansionary policy because supply is usually slow to respond to increased demand. In Zambia, supply may be inelastic because of tight capacity constraints in the availability of capital stock, executive and professional manpower, and transportation for foreign trade. Therefore, to model the process of price formation, we consider the effect of excess demand generated by changes in the money supply and the effect of cost pressures caused by labor and imported input costs.

The price deflator of aggregate output less mining output (PGDPNM) is therefore made a function of the unit of labor cost in the nonmining sector (ULCNM), the price deflator for imports of goods (PMG), and the ratio of aggregate real output in nonmining sectors to the money supply (XGDPNM65/MSQM). The resulting equation is

$$\begin{aligned} \text{PGDPNM} = \underset{(3.53)}{0.6977} &- \underset{(-2.04)}{0.0557}\ \text{XGDPNM65/MSQM} \\ &+ \underset{(2.04)}{0.2744}\ \text{PMG} + \underset{(2.38)}{0.6748}\ \text{ULCNM} \end{aligned} \qquad (4.54)$$

$$R^{-2} = 0.98 \qquad SEE = 0.031 \qquad DW = 1.58$$

The elasticities with respect to the unit labor cost and the deflator of imports of goods are 0.32 and 0.29, respectively.

Sectoral Output-Price Deflators in Nonmining Sectors. For all nonmining sectors (except agriculture), we make the price deflator for output a function of the aggregate nonmining price deflator. The wage rate in each sector was also included as an explanatory variable, but it was significant for some sectors and not for others. The estimated relationships for sectoral price deflators are as follows:

Manufacturing

$$\underset{}{\text{Ln PXMANF}} = \underset{(1.46)}{0.0216} + \underset{(38.54)}{1.4303}\ \text{Ln PGDPNM} \tag{4.55}$$

$$\bar{R}^2 = 0.993 \qquad SEE = 0.026 \qquad DW = 2.72$$

where Ln stands for the natural log.

Construction

$$\text{Ln PXCONST} = -\underset{(-3.0)}{2.0839} + \underset{(4.60)}{0.7062}\ \text{Ln PGDPNM} + \underset{(3.24)}{0.3764}\ \text{Ln WGACZ} \tag{4.56}$$

$$\bar{R}^2 = 0.926 \qquad SEE = 0.068 \qquad DW = 1.33$$

Transportation

$$\text{PTRCOM} = -\underset{(-2.60)}{0.2182} + \underset{(20.2)}{1.1478}\ \text{PGDPNM} \tag{4.57}$$

$$\bar{R}^2 = 0.974 \qquad SEE = 0.06 \qquad DW = 0.62$$

Services

$$\text{Ln PXSER} = -\underset{(-1.13)}{0.2771} + \underset{(13.1)}{0.8492}\ \text{Ln PGDPNM} + \underset{(1.1)}{0.0422}\ \text{Ln WGASZ} \tag{4.58}$$

$$\bar{R}^2 = 0.992 \qquad SEE = 0.017 \qquad DW = 2.69$$

Commercial Agriculture. The deflator for output in commercial agriculture is made a function of the wages paid in agriculture and the price of maize. This price is set for each agricultural year by the marketing board for agricultural products:

$$\begin{aligned} \text{PXAFFCS} = {} & \underset{(5.8)}{0.6944} + \underset{(2.23)}{0.0009}\ \text{WGAAFFZ} + \underset{(1.92)}{0.0622}\ \text{PPM} \\ & + \underset{(2.16)}{0.2006}\ \text{DZ71} - \underset{(-3.84)}{0.4104}\ \text{DZ75} \end{aligned} \tag{4.59}$$

$$R^{-2} = 0.725 \qquad SEE = 0.08 \qquad DW = 1.49$$

where WGAAFFZ is the average annual wage rate of Zambians in commercial agriculture, PPM is the price of maize, and DZ71 and DZ75 are dummy variables for 1971 and 1975, respectively.

Subsistence Agriculture. The price deflator depends on the price of maize. The estimated relation for PXAFFS, the deflator of output in subsistence agriculture, using 1965–1976 data, is

$$\begin{aligned} & \text{PXAFFS} = \underset{(5.26)}{0.3210} + \underset{(16.53)}{0.2217}\ \text{PPM} + \underset{(2.61)}{0.1469}\ \text{DZ73} \\ & R^{-2} = 0.962 \qquad SEE = 0.054 \qquad DW = 2.61 \end{aligned} \tag{4.60}$$

Deflator for Output in Mining and Quarrying. The deflator is dependent on the export price of copper. It is therefore made a function of the unit-value index of exports of copper.

$$\begin{aligned} & \text{PXMINQ} = -\underset{(-0.95)}{0.1729} + \underset{(11.47)}{1.2130}\ \text{PXGC} - \underset{(-9.53)}{1.1954}\ \text{DZHALF75/76} \\ & R^{-2} = 0.952 \qquad SEE = 0.1402 \qquad DW = 2.3 \end{aligned} \tag{4.61}$$

where PXMINQ is the deflator for output in mining and quarrying, PXGC is the unit-value index of exports of copper, and DZHALF75/76 is a dummy variable having the value of 0.7 in 1975 and 1.0 in 1976.

Aggregate Demand Price Deflators

There are three deflators for components of aggregate demand, for private consumption, government consumption, and gross fixed investment. Each of these deflators is made a function of the deflator for total output in nonmining sectors and import prices. The estimated equations for the deflators are as follows:

Private Consumption

$$\text{Ln PBCPC} = \underset{(0.13)}{0.0201} + \underset{(7.71)}{0.7482}\ \text{Ln PGDPNM} + \underset{(3.68)}{0.2150}\ \text{Ln PMGS} \tag{4.62}$$

$$\bar{R}^2 = 0.99 \qquad SEE = 0.0024 \qquad DW = 2.33$$

Government Consumption

$$\text{Ln PBCGC} = \underset{(-0.30)}{-0.0050} + \underset{(8.70)}{0.7768}\ \text{Ln PGDPNM} + \underset{(2.74)}{0.1473}\ \text{Ln PMGS} \tag{4.63}$$

$$\bar{R}^2 = 0.99 \qquad SEE = 0.022 \qquad DW = 2.93$$

Gross Fixed Investment

$$\text{PIGFC} = \underset{(-0.94)}{-0.1435} + \underset{(3.05)}{0.8064}\ \text{PGDPNM} + \underset{(1.90)}{0.3835}\ \text{PMMATE} \tag{4.64}$$

$$\bar{R}^2 = 0.96 \qquad SEE = 0.064 \qquad DW = 1.87$$

where PGDPNM is the price deflator for output in the nonmining sectors, PMGS is the price index of imports of goods and services, and PMMATE is the price index of imports of machinery and transportation equipment. The elasticities of the deflator for gross fixed investment (PIGFC) with respect to PGDPNM and PMMATE are 0.78 and 0.32, respectively.

Consumer Price Indexes

Two consumer price indexes are used, one for Zambians and one for expatriates. This division has its origins in preindependence days and soon afterward, when the series were for Africans and Europeans. Recently, the Zambian Central Statistical Office has replaced all the previous series with new ones. The two new definitions are price indexes for low-income groups and high-income groups. We use the indexes for Zambians and expatriates because they are better suited for our purposes. Also, all the series are highly correlated. The consumer price index for Zambians (PC) is made a function of the price deflator for output in nonmining sectors (PGDPNM), the price index for imports of goods and services (PMGS), and the implicit ratio of the money supply to the real gross national product. The price index for expatriates (VPCE) is a function of the index for Zambians and index of imports.

$$\text{Ln PC} = \underset{(49.3)}{4.1318} + \underset{(3.74)}{0.6271} \text{ Ln PGDPNM} + \underset{(3.14)}{0.1893} \text{ Ln PMGS}$$

$$+ \underset{(2.07)}{0.09294} \text{ Ln MSQM/XGNP65} \quad (4.65)$$

$$R^{-2} = 0.995 \quad SEE = 0.017 \quad DW = 1.97$$

$$\text{Ln VPCE} = \underset{(1.23)}{0.4587} + \underset{(0.41)}{0.0247} \text{ Ln PMGS} + \underset{(10.8)}{0.9827} \text{ Ln PC} \quad (4.66)$$

$$R^{-2} = 0.991 \quad SEE = 0.022 \quad DW = 1.04$$

Import-Price Indexes

There is a corresponding price index for each of the components of imports, as outlined earlier in this chapter. Import prices are taken as given. Zambia has no control over them, except that the prices paid for imports by Zambians can be affected by the exchange rate of foreign currencies, which the Zambian monetary authorities can alter. We therefore make the import-price indexes functions of the corresponding price indexes for imports of developing nations from developed economies. These series, compiled by the United Nations, are calculated on a dollar base. They are therefore converted into kwacha, using the dollar-kwacha exchange rate. The following are the estimated relationships for the period 1964–1975:

Food, Beverages, and Tobacco (SITC 0 and 1)

$$\text{Ln PM01G} = \underset{(-7.13)}{-3.2999} + \underset{(7.31)}{0.7608} \text{ Ln PXD01G/FDOLL} \quad (4.67)$$

$$R^{-2} = 0.83 \quad SEE = 0.102 \quad DW = 2.62$$

Crude Materials, Vegetable Oils, and Fats (SITC 2 and 4)

$$\text{Ln PM24G} = \underset{(-8.5)}{-5.8260} + \underset{(8.72)}{1.3649} \text{ Ln PXD24G/FDOLL} \quad (4.68)$$

$$R^{-2} = 0.87 \quad SEE = 0.118 \quad DW = 0.56$$

Mineral Fuels (SITC 3)

$$\text{Ln PM3G} = \underset{(-16.9)}{-5.4384} + \underset{(17.5)}{1.2735} \text{ Ln PXD3G/FDOLL} \quad (4.69)$$

$$R^{-2} = 0.965 \quad SEE = 0.089 \quad DW = 1.23$$

Chemicals (SITC 5)

$$\text{Ln PM5G} = -3.3595 + 0.7847 \text{ Ln PXD5G/FDOLL} \quad (4.70)$$
$$(-7.57) \quad (7.83)$$

$$\bar{R}^2 = 0.846 \quad SEE = 0.085 \quad DW = 1.41$$

Machinery and Transportation Equipment (SITC 7)

$$\text{Ln PMMATE} = -4.376 + 1.0331 \text{ Ln PXD7G/FDOLL} \quad (4.71)$$
$$(-8.14) \quad (8.40)$$

$$\bar{R}^2 = 0.864 \quad SEE = 0.098 \quad DW = 1.17$$

Manufactured Goods (SITC 6 and 8)

$$\text{Ln PMMAG} = -5.2593 + 1.2505 \text{ Ln PXD68G/FDOLL} \quad (4.72)$$
$$(-14.9) \quad (15.40)$$

$$\bar{R}^2 = 0.956 \quad SEE = 0.062 \quad DW = 1.28$$

Miscellaneous Manufactures (SITC 6 and 8)

$$\text{Ln PMMMA} = -4.1341 + 0.97884 \text{ Ln PXD68G/FDOLL} \quad (4.73)$$
$$(-7.52) \quad (7.74)$$

$$\bar{R}^2 = 0.843 \quad SEE = 0.096 \quad DW = 1.76$$

Services

$$\text{PMSS} = -5.1051 + 3.930 \text{ FDOLL} + 0.0366 \text{ TIME} \quad (4.74)$$
$$(-3.38) \quad (2.97) \quad (1.31)$$

$$\bar{R}^2 = 0.801 \quad SEE = 0.184 \quad DW = 1.75$$

PXD01G is the unit-value index of goods exported to developing areas for SITC categories 0 and 1, expressed on a dollar base; PXD24G, PXD3G, and so forth, are defined accordingly; and FDOLL is the exchange rate in dollars per kwacha.

The coefficients of the unit-value indexes of foreign trade may reflect the tariff structure. The coefficient for machinery and equipment is close to 1.0, since a low tariff was imposed on those goods. Until recently (1976) foods were not subject to tariffs, and food imports were sometimes subsidized. This also applies to chemicals, especially fertilizer.

Unit-Value Index of Copper Exports

This index is simply expressed as a function of the export price of copper in kwacha per metric ton. The resulting equation is

$$\underset{}{PXGC} = \underset{(1.45)}{0.0069} + \underset{(304.1)}{0.0020}\ EXCPRICE \qquad (4.75)$$

$$\bar{R}^2 = 1.00 \qquad SEE = 0.0075 \qquad DW = 0.87$$

Government Revenues and Expenditures

This section deals with recurrent government revenues and expenditures. Only the central-government fiscal environment is considered; there are no data for a host of provincial and local authorities that also levy taxes and spend money to provide services. It is estimated, however, that central-government revenues and expenditures account for over 90 percent of total government finances. Therefore, we are capturing the main ingredients of public financing.

Revenues

We explain revenues by source. Government revenues come from customs duties, excise and sales taxes, income taxes, mineral revenues, and other revenues. The other-revenues category consists of such miscellaneous items as licenses, fines, and court fees.

Nonmineral Revenues. We estimated equations for customs duties (CGCD), excise and sales taxes (CGEST), income taxes (CGIT), and other revenues (CGRO) as functions of relevant income or economic activity. CGCD is a function of the dutiable portion of imports of goods and services (MGTT). This is merchandise imports, less chemicals and machinery and transportation equipment. CGEST is made a function of gross domestic output less exports of goods and services. Income taxes (CGIT) are made a function of wage income and nonmining operating surplus, less output in subsistence agriculture. Other revenues (CGRO) are dependent on gross domestic output (XGDMC). The estimated equations are as follows:

Customs Duties

$$CGCD = \underset{(-0.70)}{-6.6488} + \underset{(4.29)}{0.1998}\ MGTT + \underset{(1.73)}{16.2131}\ DZ72$$

$$\bar{R}^2 = 0.645 \qquad SEE = 8.96 \qquad DW = 1.14 \qquad (4.76)$$

Excise and Sales Taxes

$$CGEST = -106.42 + 0.2326\ (XGDMC - EXGS) \tag{4.77}$$
$$(-4.3) \quad (6.89)$$

$$\bar{R}^2 = 0.81 \quad SEE = 23.8 \quad DW = 0.81$$

Income Taxes

$$CGIT = -63.735 + 0.3169\ YWC + 0.0387\ PROF + 38.2281\ DZ76 \tag{4.78}$$
$$(-7.24) \quad (15.1) \quad (1.37) \quad (3.89)$$

$$\bar{R}^2 = 0.97 \quad SEE = 7.96 \quad DW = 1.81$$

Other Revenues

$$CGRO = 4.8187 + 0.0294\ XGDMC + 21.6752\ DZ72 \tag{4.79}$$
$$(1.43) \quad (11.6) \quad (5.95)$$

$$\bar{R}^2 = 0.936 \quad SEE = 3.5 \quad DW = 2.15$$

where DZ72 is a dummy variable (1.0 in 1972, 0 otherwise); EXGS is exports of goods and services; YWC is total compensation of employees, less employee benefits; PROF is total operating surplus, less mining profits and output of subsistence agriculture; and DZ76 is a dummy variable equal to 1.0 in 1976 and 0 otherwise. It represents the effect of tax increases in 1976.

Mineral Revenues. The mineral revenues are obtained by applying the tax rates for the various kinds of taxation that have been in force since independence to output and earnings in the copper sector. Until 1966, copper companies paid two kinds of taxes to the government. One form was royalties, which was based on the London Metal Exchange (LME) price of copper; the other was company tax, based on profitability. In 1966, the government introduced the export tax on copper, based on the LME price. In 1970, the government abolished royalty payments and export taxes and replaced them with mineral taxes based on profitability.

Royalty Payment. Royalties were calculated on a formula that was 13.5 percent of the price of copper, less sixteen kwacha per long ton of the copper produced. The calculation of the price used a complicated combination of monthly average prices for various types of copper on the LME. However, for simplicity, we used the LME price of copper wirebars (PCLME). Hence, for the period when it was applicable (up to 1969), royalty payments were given by

$$VTAXROY = (0.135 * PCLME/EXCH - 16.0) * CUQCZM/1.10231 \tag{4.80}$$

where CUQCZM is the copper output in metric tons and EXCH is the exchange rate in pounds sterling per kwacha.

Export Tax. The export tax was imposed to take advantage of the rising LME copper prices in the early years of independence. It was instituted in 1966 and abolished in 1969. The export tax was charged on 40 percent of the price of copper per long ton in excess of 600 kwacha. It is calculated as

$$VTAXEX = 0.4 * (PCLME/EXCH - 600) * EXCOPQT/1.10231 \quad (4.81)$$

where EXCOPQT is copper in metric tons.

Mineral Tax. The mineral tax replaced the export-tax and royalty system in 1970. It is applied on gross profits, less allowance for capital expenditures. The reforms of 1970 allowed companies to deduct net capital expenditures from profits in the year they were made, for tax purposes. The tax base for mineral and income tax is therefore given by

$$VGTAXB = VGP - VCAARATE * KCMIN \quad (4.82)$$

where VGP is gross profits, VCARRATE is the fraction of capital expenditures that can be written off, and KCMIN is the amount of capital expenditures. The mineral revenues are calculated as a percentage of the tax base. The applicable rate at the time of introduction was 51 percent. The mineral tax is given by

$$VMTAX = VMTAXRATE * VGTAXB \quad (4.83)$$

where VMTAXRATE is the rate of mineral tax.

Company Taxes. Company taxes apply to all firms doing business in Zambia. They are calculated as a percentage of taxable profits. For the mining companies, taxable profits for company-tax purposes are the tax base (VGTAXB), less mineral taxes, export taxes, and royalties, whenever these taxes are applicable. Therefore, company tax is calculated as

$$VTAXCOMP = VCTAXRATE * (VGTAXB - VTAXES - VTAXROY - VMTAX) \quad (4.84)$$

where VCTAXRATE is the rate of company taxation. In the past, this rate varied from 45 percent to 50 percent.

Total Mineral Revenues. Total mineral revenues are calculated as the sum of different taxes:

$$\text{CGMR} = \text{VTAXCOMP} + \text{VTAXEX} + \text{VTAXROY} + \text{VTMAX} \tag{4.85}$$

where VTAXCOMP applies to all periods; VTAXEX applies only to 1966–1969 and is zero for other periods; VTAXROY applies only to 1964–1969 and zero otherwise; and VMTAX applies to 1970 on.

Government Expenditures

Three components of government current expenditures are explained: expenditures on general services, social services, and economic services. Expenditures on social services (CGSS) go mainly to education and health. Economic-services expenditures (CGES) are current expenditures for the promotion of economic activities, for example, services to agriculture and rural development, transportation, communications, and power. Expenditures on general services (CGGS) go for the provision of administrative services by government departments.

In the explanation of expenditures, we emphasize that the role of revenues lagged one period. The amount of revenues received in this period may influence expectations of revenues in the next period, or it determines the amount of surplus or deficit revenues with which the government can work next period. Thus, lagged government revenues enter into our explanation of the three components of expenditures. In addition, the urban population (POPU) enters into the explanation of CGGS, the total population (POP) is used for CGSS, and the gross domestic product (XGDMC) enters into the explanation of CGES. The estimated equations for the 1966–1976 period are as follows:

General Services

$$\underset{}{\text{CGGS}} = \underset{(-5.1)}{-\,104.20} + \underset{(5.76)}{129.935}\ \text{POPU} + \underset{(2.68)}{0.1446}\ \text{CGRR}(-1)$$

$$R^{-2} = 0.943 \qquad SEE = 14.09 \qquad DW = 1.44 \tag{4.86}$$

Social Services

$$\text{CGSS} = \underset{(-8.3)}{-\,201.211} + \underset{(9.12)}{62.297}\ \text{POP} + \underset{(0.65)}{0.0150}\ \text{CGRR}(-1)$$

$$R^{-2} = 0.957 \qquad SEE = 6.27 \qquad DW = 0.82 \tag{4.87}$$

Economic Services

$$\text{CGES} = \underset{(-1.2)}{-\,21.2355} + \underset{(1.97)}{0.315}\ \text{XGDMC} + \underset{(5.13)}{0.1913}\ \text{CGRR}(-1)$$

$$R^{-2} = 0.859 \qquad SEE = 12.8 \qquad DW = 2.25 \tag{4.88}$$

where CGRR is the total recurrent revenue of the government. Lagged government revenues are not significant for expenditures on social services, because education and health have always received high priority and have received the necessary funds irrespective of revenue conditions. The variable for total population (POP) captures the fact that expenditures on social services have been increasing along a trend. Lagged revenues are most significant for economic services, indicating that this is the most discretionary component for the allocation of funds and other nonmonetary resources.

Financial Flows

This section deals with domestic and international financial flows. The domestic flows explained are the change in government debt and the change in money supply. The international financial flows explained include personal transfers, investment transfers, and international reserves.

Net Claims on the Government

The change in net claims on the government (DNNGD) is made a function of the net surplus or deficit in government activities, including government capital expenditures. It is posited that the government finances its deficit by selling bonds and treasury bills, hence increasing its indebtedness, and uses its surplus to reduce this debt by buying back some of its bills and bonds. The estimated relation for the period 1966–1976 is

$$\underset{(-1.63)\quad(-10.1)}{\text{DNNGD} = -\,23.047 - 0.9058\ \text{CGND}}$$

$$\bar{R}^2 = 0.91 \qquad SEE = 38.06 \qquad DW = 1.34 \tag{4.89}$$

where CGND is the net government surplus (+) or deficit (−). The negative sign on CGND indicates that, if there is a surplus, DNNGD goes down, and the reverse for a deficit. We obtain the net claims on the government as follows:

$$\text{NNGD} = \text{NNGD}(-1) + \text{DNNGD} \tag{4.90}$$

Money Supply

The supply of money changes in response to government borrowing from the Central Bank and commercial banks. When the government borrows from the

Central Bank and spends the money, the monetary base changes and the money supply may change. When the government borrows from the commercial banks, money supply may or may not change. In Zambia, government securities form part of the banks' reserve assets. Hence, borrowing by the government to finance deficits is likely to lead to changes in the money supply, and the money supply will be sensitive to government deficit or surplus. Another factor in changes in the money supply is the change in foreign assets. If Zambia's holdings of foreign assets rise, the domestic money supply will be affected when these foreign assets are converted to local currency by the Central Bank.

The change in the money supply (DMSQM), therefore, is modeled as a function of the change in the net claims on the government (DNNGD) and the change in net foreign assets (BPNECF). The resulting equation obtained for the period 1966–1976 is

$$\begin{aligned} \text{DMSQM} = \underset{(5.76)}{35.8488} + \underset{(4.06)}{0.3618}\ \text{BPNECF} + \underset{(2.46)}{0.1707}\ \text{DNNGD} \\ + \underset{(3.43)}{63.2102}\ \text{DZ73} \end{aligned} \qquad (4.91)$$

$$R^{-2} = 0.739 \qquad SEE = 17.54 \qquad DW = 3.04$$

where DZ73 is a dummy variable for 1973. It represents the easy credit conditions that were in effect in that year. Having obtained the change in money supply, we obtain money supply as

$$\text{MSQM} = \text{MSWM}(-1) + \text{DMSQM} \qquad (4.92)$$

where MSQM stands for money supply broadly defined, that is, currency in the hands of the public plus demand and time deposits. Activities of the copper industry can affect the money supply through both the net claims on the government, and the change in net foreign assets. If copper prices rise, BPNECF rises, government revenues increase and government deficits should decrease. Thus, the two variables act in offsetting directions.

International Financial Flows

Investment income paid abroad depends on profits in the Zambian economy. Net noninvestment income paid abroad depends on the earnings of expatriates in Zambia, while investment income earned abroad depends on Zambian holdings of foreign assets. Net international-reserve position depends on the lagged value of reserves and the balance of capital and current accounts.

Investment Income Paid Abroad

$$\text{YIEX} = \underset{(24.39)}{68.5633} + \underset{(1.35)}{0.0313}\ \text{YPROFR} - \underset{(-4.22)}{22.8837}\ \text{DZ70/71} + \underset{(4.53)}{19.2144}\ \text{DZ70/76} - \underset{(-5.27)}{37.6172}\ \text{DZ75} \qquad (4.93)$$

$R^{-2} = 0.766$ $\quad SEE = 6.25 \quad DW = 2.02$

where YPROFR is total operating surplus, less output in subsistence agriculture, mineral taxes, and indirect taxes; and DZ70/71, DZ70/76, DZ75 are dummy variables for 1970–1971, 1970–1976 and 1975, respectively. The dummy variable DZ70/71 may be the effect of restrictions on dividends that were put into place in 1969. DZ70/76 reflects the effect of compensation payments arising from the takeover of the copper industry, and DZ75 represents strict control on foreign exchange that was in force in a year of scarce foreign exchange.

Noninvestment Income Paid Abroad (−)

$$\text{YNOEX} = -\underset{(-7.0)}{0.3384}\ \text{YWEXPR} - \underset{(-4.7)}{0.1820}\ \text{NINTR}(-1) + \underset{(5.78)}{42.2777}\ \text{DZ51/69} + \underset{(2.29)}{29.603}\ \text{DZ67} \qquad (4.94)$$

$R^{-2} = 0.913 \quad SEE = 11.17 \quad DW = 1.51$

where YWEXPR is the total wage compensation of expatriates, NINTR is the net total international reserves, DZ51/69 is a dummy variable for the precopper-industry-nationalization period, and DZ67 is a dummy variable for 1967. The variable NINTR indicates the willingness and ability to allow for transfer of incomes. When reserves are low, less income is being transferred because either more stringent controls are in force or the existing controls are enforced more strictly. The coefficient of the dummy variable DZ51/69 indicates that transfers abroad increased after the takeover. The equation was estimated in the 1966–1976 period.

Investment Income from Abroad

$$\text{YIM} = \underset{(4.62)}{10.2416} + \underset{(3.71)}{0.0457}\ \text{NFORA} \qquad (4.95)$$

$R^{-2} = 0.538 \quad SEE = 5.58 \quad DW = 2.35$

where NFORA is the net holdings of foreign assets.

Net International Reserves

$$\underset{}{\text{NINTR}} = \underset{(-0.64)}{-32.6858} + \underset{(4.25)}{1.00036}\ \text{NINTR}(-1) + \underset{(3.23)}{0.7621}\ \text{BPC} + \underset{(2.18)}{0.7100}\ \text{BPCA} \qquad (4.96)$$

$R^{-2} = 0.707 \quad SEE = 48.0 \quad DW = 2.08$

where BPC is balance of current account and BPCA is the balance of capital account. The relationship was obtained for the 1967–1976 period.

Miscellaneous Variables

To close the model, we present estimates for consumption of fixed capital, indirect taxes, and urban population. The consumption of fixed capital is a function of total capital stocks. To obtain a nominal value of capital stocks, we simply multiplied the current price deflator for fixed investment (PIGFC) by the real capital stocks (KS65). Indirect taxes are made a function of excise and sales taxes and mineral taxes. It is assumed that a portion of these taxes is classified as indirect taxes. Urban population follows the Harris-Todaro (1970) rural-urban migration model. It makes the ratio of urban population to total population (POPU/POP) a function of the real wages paid in mining (WGAMQ/PC) and a time trend (TIME). Since mining wages determine wages in other urban sectors of the economy, they are used as a proxy for urban wages. The estimated equations are as follows:

Consumption of Fixed Capital

$$\text{CFC} = \underset{(.71)}{8.6992} + \underset{(13.89)}{0.0450}\ \text{KS65} * \text{PIGFC} \qquad (4.97)$$

$R^{-2} = 0.941 \quad SEE = 20.5 \quad DW = 0.4$

Indirect Taxes

$$\text{INDTAX} = \underset{(1.35)}{51.8755} + \underset{(2.66)}{0.7833}\ \text{CGEST} + \underset{(3.97)}{0.6932}\ \text{CGMR} \qquad (4.98)$$

$R^{-2} = 0.581 \quad SEE = 49.1 \quad DW = 2.10$

Urban Population

$$\text{POPU/POP} = \underset{(2.74)}{0.0178} + \underset{(4.51)}{0.001322}\ \text{WGAMQZ/PC} + \underset{(55.5)}{0.01279}\ \text{TIME} \qquad (4.99)$$

$R^{-2} = 0.996 \quad SEE = 0.00345 \quad DW = 1.35$

where CGEST is excise and sales taxes; CGMR is mineral taxes; POPU is urban population; POP is total population; and PC is the consumer price index. The estimation periods for equations 4.97–4.99 are 1964–1976, 1965–1976 and 1963–1976 respectively.

Notes

1. See *A Survey of Zambian Industry*. Lusaka, Indeco, Ltd. 1972.
2. See Arrow (1974).
3. Republic of Zambia, *Census of Agriculture 1970–1971 (Second Report);* Central Statistical Office, Lusaka, June 1977.
4. See Bodkin (1977).
5. See Jorgenson (1969).
6. See Jorgenson (1971).

References

Arrow, K.J. 1974. "Measurement of Real Value Added." In P. David and M. Reder (eds.), *Nations and Households Economic Growth*. New York: Academic Press.

Barro, R., and Grossman, H. 1976. *Money, Employment and Inflation*. Cambridge, England: Cambridge University Press.

Bodkin, R.G. 1977. "Keynesian Econometric Concepts: Consumption Function, Investment Function and 'The' Multiplier." In S. Weintraub (ed.), *Modern Economic Thought*. Philadelphia: University of Pennsylvania Press.

Clower, R. 1965. "The Keynesian Counter-Revolution; A Theoretical Appraisal." In F.H. Hahn and F.P.R. Brechling (eds.), *The Theory of Interest Rates*. London: Macmillan.

Harris, J., and Todaro, M. 1970. "Migration, Unemployment and Development: A Two-Sector Analysis." *American Economic Review* 60:126–142.

Jorgenson, D.J. 1969. "The Theory of Investment Behavior." In H. Williams and J. Huffnagle (eds.), *Macroeconomic Theory: Selected Readings*. Englewood Cliffs, N.J.: Prentice-Hall.

——— 1971. "Econometric Studies of Investment Behavior: A Survey." *Journal of Economic Literature* 9: 1111–1147.

Model Validation, Multiplier Analysis, and Copper-Price-Fluctuation Tests

The test of a macroeconometric model is how well it represents the actual economy. The model should be able to predict the endogenous variables fairly accurately, that is, within reasonable limits of error. In addition, the model should display sensitivity to changes in the exogenous variables, which can be explained by economic theory, by empirical observation, or by both. We shall assess the performance of the model of Zambia by testing its forecasting ability and its sensitivity to exogenous shocks.

In chapter 3, alternative models of the output decision-making process in the copper industry were presented. One model assumes that output is determined based on a profit-maximizing objective. The second model assumes that the industry combines available capital stock with labor to maximize output. Also in chapter 3, two models of the investment-decision process were presented. One is an expectations model, which makes future investment a function of past prices of copper; the second model makes investment a function of lagged prices and profits. In the historical simulation and multiplier analysis, we use the profit-maximizing output decision model for copper production and the profit-prices-investment model, unless otherwise stated. The justification is that the profit-maximizing output model and the profit-prices-investment model probably correspond more to historical reality than the alternatives.

Dynamic Historical Simulation, 1968-1976

The predictive ability of the model was assessed by comparing dynamic-historical-period solution values with the actual values of the endogenous variables. A dynamic solution is a historical-period simulation that uses actual values for the iterative process in each period, solution values for the lagged values in the model, and actual values of the exogenous variables. In contrast, a forecast solution uses one-period lags of solution values as starting values.

To compare the dynamic-solution values with the actual values, we make use of the following summary statistics: mean absolute percentage deviation (MAPE), which is given by

$$\text{MAPE} = \frac{1}{T} \sum_{t=1}^{T} 100.0 * \left| (x_s^t - X_a^t)/X_a^t \right|$$

where T is the period of the solution, X_s^t is the solution value of the endogenous variable at time t, and X_a^t is the actual value of the endogenous variable; and mean of actuals and mean of predicteds, where

$$\text{Mean of actuals} = \frac{1}{T} \sum_{t=1}^{T} X_a^t$$

and similarly for mean of predicteds.

In table 5–1, we present these statistics for a set of selected major macroeconomic variables. In figure 5–1, we present graphical comparisons of historical-solution values and actual values for some variables. These graphs give an indication of how well the model captures the turning points of the variables.

There is no standard benchmark of performance against which the calculated summary statistics are to be judged. However, in assessing the performance of the model, two considerations are relevant. The first is how the model performs in comparison to other models of similar size, type, and class; the second pertains to the uses for which the model is intended.

Priovolos compared the predictive performance of six econometric models of five so-called developing nations. There was one model each for Ivory Coast, Zambia, Chile, and Brazil and two for Mexico.[1] The measure of performance was the mean absolute percentage deviation for several key variables for each model. The Zambian model performed very well when judged against the other models.

A model that is intended to be used mainly for simulation exercises need not have the kind of predictive performance that is usually desired in forecasting models. In fact, forecasting ability may sometimes be sacrificed to obtain better simulation properties. For example, in our work, we could have exogenized

Table 5-1
Error Statistics, Means of Actuals, and Means of Predicteds of Selected Variables

Variables	*Mean Absolute Percentage Deviation*	*Mean of Actuals*	*Mean of Predicteds*
GDP, real	3.30	896.00	885.70
Government-consumption expenditure	6.10	300.00	300.20
Government current expenditure	6.47	358.80	354.70
Government capital expenditure	6.81	174.90	168.40
Government current revenue	5.20	409.50	397.00
Copper production (tonnes)	2.06	694.30	695.80
Total employment	2.10	360.50	359.10
Final demand, real	4.02	841.60	845.50
Money supply	7.20	403.10	434.10
Consumer price index	3.81	83.70	82.40
GDP price deflator	2.63	1.63	1.62
Urban population	1.23	1.47	1.46
Exports of goods and services, real	3.31	401.80	409.30
Imports of goods and services, real	8.64	375.30	387.30
Balance of trade	19.80	256.40	256.70

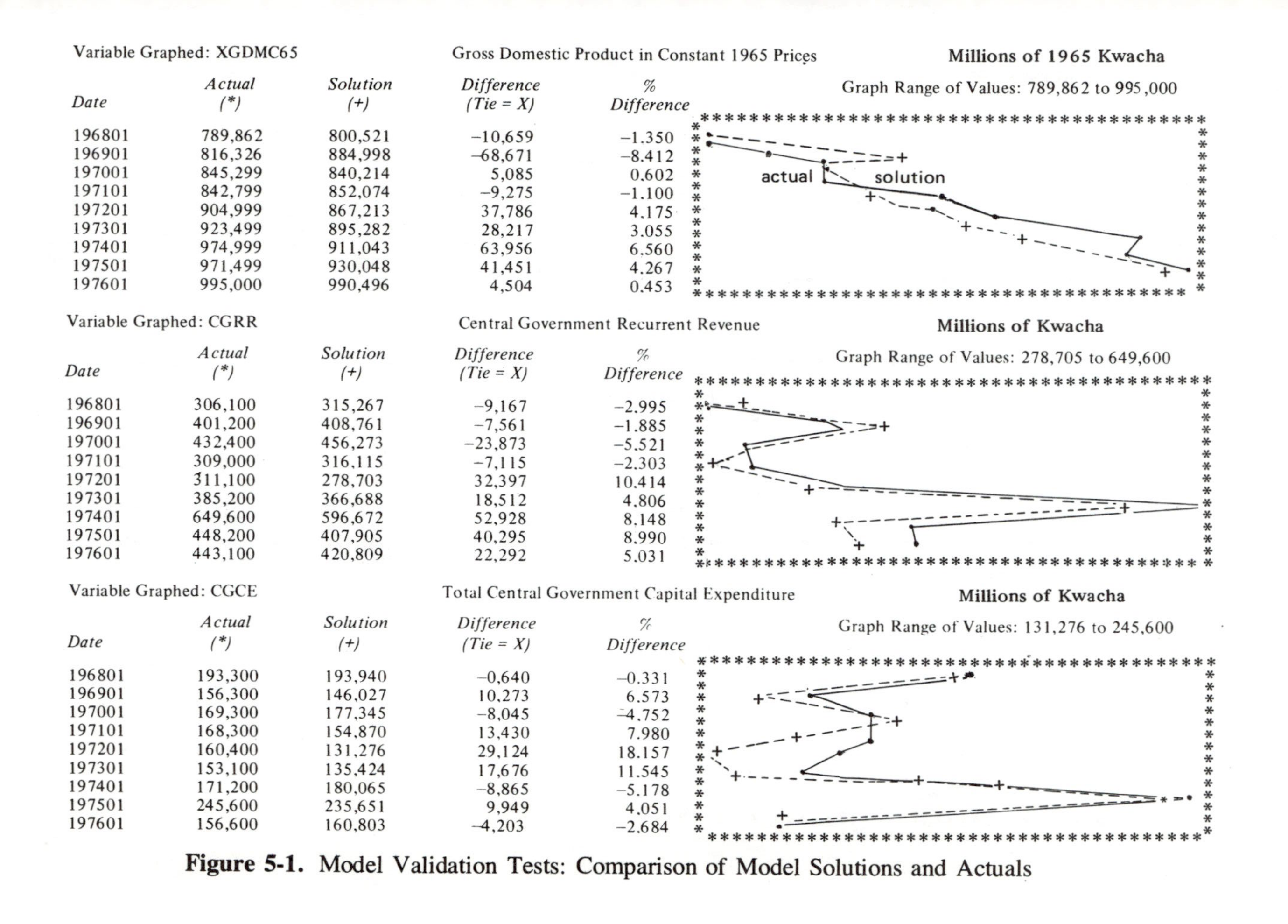

Variable Graphed: XGDMC65 — Gross Domestic Product in Constant 1965 Prices — **Millions of 1965 Kwacha**

Date	Actual (*)	Solution (+)	Difference (Tie = X)	% Difference
196801	789,862	800,521	−10,659	−1.350
196901	816,326	884,998	−68,671	−8.412
197001	845,299	840,214	5,085	0.602
197101	842,799	852,074	−9,275	−1.100
197201	904,999	867,213	37,786	4.175
197301	923,499	895,282	28,217	3.055
197401	974,999	911,043	63,956	6.560
197501	971,499	930,048	41,451	4.267
197601	995,000	990,496	4,504	0.453

Variable Graphed: CGRR — Central Government Recurrent Revenue — **Millions of Kwacha**

Date	Actual (*)	Solution (+)	Difference (Tie = X)	% Difference
196801	306,100	315,267	−9,167	−2.995
196901	401,200	408,761	−7,561	−1.885
197001	432,400	456,273	−23,873	−5.521
197101	309,000	316,115	−7,115	−2.303
197201	311,100	278,703	32,397	10.414
197301	385,200	366,688	18,512	4.806
197401	649,600	596,672	52,928	8.148
197501	448,200	407,905	40,295	8.990
197601	443,100	420,809	22,292	5.031

Variable Graphed: CGCE — Total Central Government Capital Expenditure — **Millions of Kwacha**

Date	Actual (*)	Solution (+)	Difference (Tie = X)	% Difference
196801	193,300	193,940	−0,640	−0.331
196901	156,300	146,027	10,273	6.573
197001	169,300	177,345	−8,045	−4.752
197101	168,300	154,870	13,430	7.980
197201	160,400	131,276	29,124	18.157
197301	153,100	135,424	17,676	11.545
197401	171,200	180,065	−8,865	−5.178
197501	245,600	235,651	9,949	4.051
197601	156,600	160,803	−4,203	−2.684

Figure 5-1. Model Validation Tests: Comparison of Model Solutions and Actuals

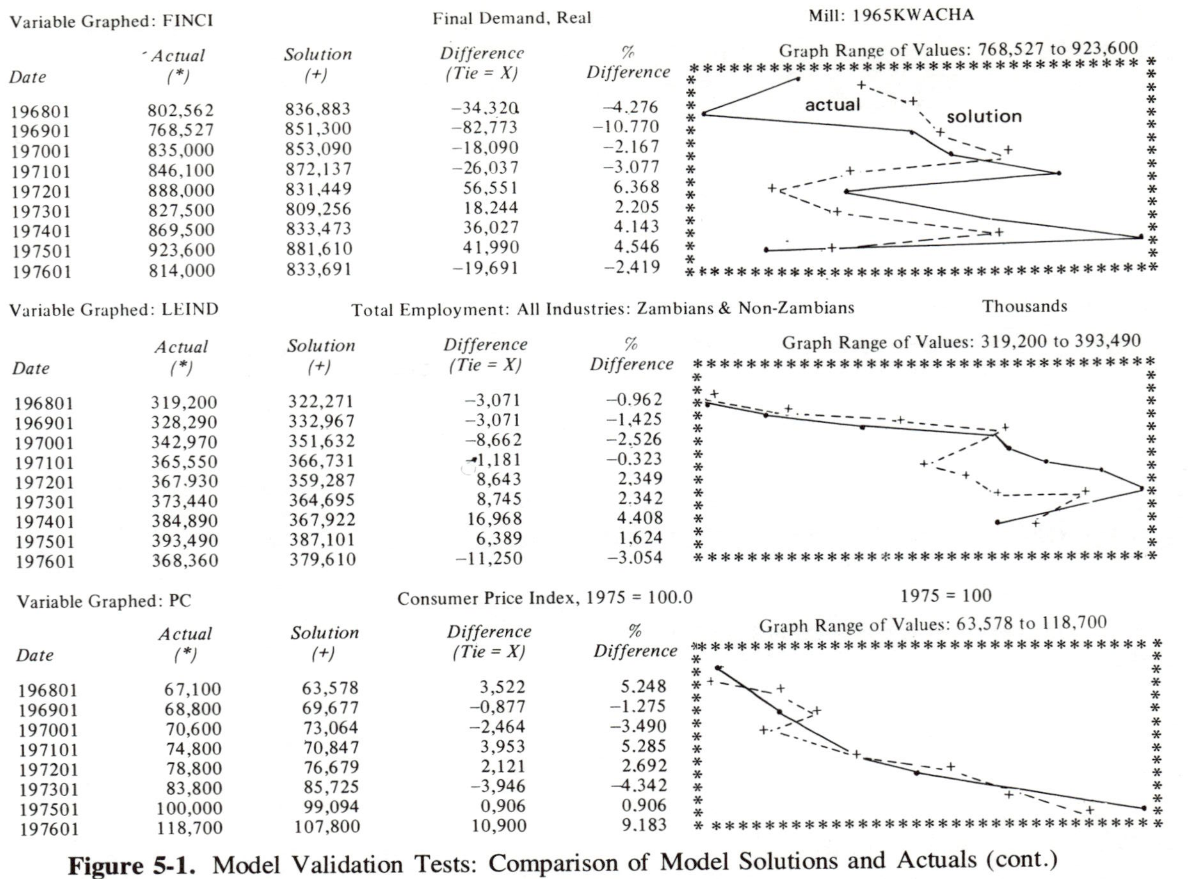

Variable Graphed: FINCI — Final Demand, Real — Mill: 1965KWACHA

Graph Range of Values: 768,527 to 923,600

Date	Actual (*)	Solution (+)	Difference (Tie = X)	% Difference
196801	802,562	836,883	−34,320	−4.276
196901	768,527	851,300	−82,773	−10.770
197001	835,000	853,090	−18,090	−2.167
197101	846,100	872,137	−26,037	−3.077
197201	888,000	831,449	56,551	6.368
197301	827,500	809,256	18,244	2.205
197401	869,500	833,473	36,027	4.143
197501	923,600	881,610	41,990	4.546
197601	814,000	833,691	−19,691	−2,419

Variable Graphed: LEIND — Total Employment: All Industries: Zambians & Non-Zambians — Thousands

Graph Range of Values: 319,200 to 393,490

Date	Actual (*)	Solution (+)	Difference (Tie = X)	% Difference
196801	319,200	322,271	−3,071	−0.962
196901	328,290	332,967	−3,071	−1,425
197001	342,970	351,632	−8,662	−2,526
197101	365,550	366,731	−1,181	−0.323
197201	367.930	359,287	8,643	2,349
197301	373,440	364,695	8,745	2.342
197401	384,890	367,922	16,968	4.408
197501	393,490	387,101	6,389	1,624
197601	368,360	379,610	−11,250	−3.054

Variable Graphed: PC — Consumer Price Index, 1975 = 100.0 — 1975 = 100

Graph Range of Values: 63,578 to 118,700

Date	Actual (*)	Solution (+)	Difference (Tie = X)	% Difference
196801	67,100	63,578	3,522	5.248
196901	68,800	69,677	−0,877	−1.275
197001	70,600	73,064	−2,464	−3.490
197101	74,800	70,847	3,953	5.285
197201	78,800	76,679	2,121	2.692
197301	83,800	85,725	−3,946	−4.342
197501	100,000	99,094	0,906	0.906
197601	118,700	107,800	10,900	9.183

Figure 5-1. Model Validation Tests: Comparison of Model Solutions and Actuals (cont.)

government expenditures (as is the practice in many forecasting models) to get better forecasts. However, since our primary interest is in simulation work, an endogenous government serves this end better.

The comparison tests indicate that the variables that show the greatest deviation from actuals are those that usually exhibit high year to year variations. The variables for the balance of current account (BPC), the balance of trade (BPT), the government spending deficit or surplus (CGND), the net international reserves (NINTR), and the gross profits of the copper industry (VGP) show high deviations from actuals, as indicated by the mean percentage error values. Most of the other variables have percentage deviations much lower than 10 percent. On the basis of these results, we conclude that the model performed well.

Multiplier Analysis

The model is tested for its response to exogenously imposed shocks. The dynamic-historical-period solution for 1968-1976 is used as a base solution. Shocks can be imposed by changing the value of an exogenous variable, with other exogenous variables keeping the same values as in the base solution. The model is solved and the solution values are compared with the values of the base solution to obtain the impact of the shock. Endogenous variables can also be shocked, which can be done in a variety of ways. The endogenous variable could be exogenized and given values different from its values in the base solution. Alternatively, a shock could be imposed on an endogenous variable by changing the values of the coefficients in its equation. For example, we might be interested in the effect of a change in the marginal propensity to consume out of wage income. By changing the coefficient of the wage income in the consumption function, the desired test can be performed. Or we could test the effect of a change in the average propensity to consume out of all income by changing the constant term in the consumption function. Two types of shocks can be applied: a one-period shock, in which the shock is applied for only one year of the solution interval, and a sustained shock, which is applied for all years in the solution interval.

To analyze the effect of the shocks, dynamic multipliers are calculated. For sustained shocks, a multiplier value is calculated for each period. It is given by

$$M_t = (X_s^t - X_c^t)/\Delta Z^t$$

where X_s^t and X_c^t are the values of endogenous variables X in the shocked and base solutions, respectively, and ΔZ_t is the shock applied. In general, ΔZ_t will, not be equal to $Z_s^t - Z_c^t$ if Z is an endogenous variable. Hence, the multiplier effect will be given by where

$$Z_s^t - Z_c^t = \Delta Z^t + \Delta^2 Z^t$$

and $\Delta^2 Z^t$ is the induced effects of exogenous changes $\Delta Z^t, \Delta Z^{t-1}, \Delta Z^{t-2}, \ldots, \Delta Z^1$.

For the one-period shock, the dynamic multiplier is calculated as

$$M_t = \sum_{t=1}^{T} (X_s^t - X_c^t)/\Delta Z$$

where ΔZ is the exogenous shock applied in the period $t=1$ and T is the length of the simulation period.

In order to analyze the distributional effects of some exogenous changes, the marginal share of gross domestic output may be calculated: It is given by

$$MS_t = 100.0 * (X_s^t - X_c^t)/(y_s^t - y_c^t)$$

where y_s^t and y_c^t are the gross domestic product for shocked and base solutions, respectively, and $y_s^t - y_c^t$ is the induced effect of the exogenous change in Z on the gross domestic product. The marginal share can be compared with the average share to determine the impact of the exogenous change on the distribution of income. Average share is given by

$$AS_t = 100.0 * X_c^t / y_c^t$$

The effect of exogenous changes can also be analyzed by the use of some summary statistics. In addition to the mean absolute percentage deviation (MAPE), which was defined earlier, we obtained the following other statistics: mean absolute deviation (MAE) is given by

$$MAE = \frac{1}{T} \sum_{t=1}^{T} | X_s^t - X_c^t |$$

root mean squared deviation (RMSE) is given by

$$RMSE = \left[\frac{1}{T} \sum_{t=1}^{T} (X_s^t - X_c^t)^2 \right]^{1/2}$$

and root mean squared percentage deviation (RMSPE) is given by

$$RMPSE = \left[\frac{1}{T} \sum_{t=1}^{T} \quad 100.0\,(X_s^t - X_c^t)/X_c^t \quad ^2 \right]^{1/2}$$

where T is the length of the simulation period.

We shall present multiplier calculations and summary statistics for a small set of major microeconomic variables and graphical comparison of shocked and base solutions for key variables for some simulations.

The Effect of Copper Prices

The London Metal Exchange price of copper is exogenous in the model. Zambian export price of copper is based on this price and thus it has an effect on the economic activity in the Zambian copper industry and macroeconomy. Since our primary focus is on the role of copper in the economy, we studied the effects of changes in the LME price of copper in great detail. The following copper-price-related multiplier tests are performed with the model:

1. A one-period increase in the LME price of copper, with prices held the same as in the base solution in other periods.
2. A one-period decrease in the LME price of copper.
3. A sustained increase in the price of copper throughout the solution period. This increase preserves the pattern of price fluctuations for the period.
4. A sustained fall in the price of copper, with the pattern of fluctuations preserved.

We shall present and analyze the results of these simulation tests separately.

A 10 Percent Increase in the LME Price of Copper in 1968. The LME price of copper was increased by 10 percent in 1968, and the model was solved from 1968 to 1976. The values from this solution are compared with those of the base solution.

Table 5–2 shows multiplier calculations and year-by-year differences of shocked and base-solution values and table 5–3 shows the percentage differences for selected major variables. In figure 5–2, graphical comparisons of the shocked and base solutions are presented for a few major variables.

In table 5–2, the multipliers represent the effect on the selected variables for a 100 kwacha per metric ton increase in copper prices. For example, the nominal gross domestic product increases by 114.8 million kwacha per 100 kwacha increase in copper price per metric ton. The result indicates that increasing copper prices have a stimulative effect on the Zambian economy, as indicated by the increases in real gross domestic product, investment, and employment. However, the price level increased significantly and the nominal dynamic multiplier for gross domestic product is 114.8, while that for real product was only 21.8. The major real effect of the shock came in the year after the shock but declined from then so that, by 1976, the effect was insignificant. The inflationary effect of the

Table 5-2
Multipliers and Differences between Shocked and Base Solutions for a 10 Percent Increase in Copper Prices in 1968

		Differences in Solutions' Values								
Variables	*Dynamic Multipliers*	*1968*	*1969*	*1970*	*1971*	*1972*	*1973*	*1974*	*1975*	*1976*
GDP	114.80	63.30	16.80	8.30	2.10	1.80	2.30	2.20	1.80	1.50
GDP, real	21.80	4.30	5.80	3.20	1.70	1.00	0.40	0.82	0.56	0.74
Government-consumption expenditure	6.30	0.00	6.10	0.79	–0.76	–0.40	–0.17	–0.05	–0.02	–0.03
Government current expenditure	21.30	2.10	16.90	2.50	–1.80	–0.88	–0.31	0.02	0.05	0.04
Government capital expenditure	17.20	3.20	15.30	–1.20	–2.10	–0.41	–0.60	0.12	0.06	0.07
Government current revenue	52.80	45.00	6.80	–4.40	–2.10	–0.72	0.17	0.14	0.27	–0.08
Final demand, real	36.80	15.60	22.10	6.20	–2.40	–2.50	–2.10	–1.70	–1.70	–1.50
Gross fixed investment, real	36.50	6.10	14.90	7.20	1.30	0.81	0.54	0.46	0.25	0.23
Money supply	73.20	13.10	13.40	10.30	6.30	5.40	4.80	4.20	3.50	2.90
Consumer price index (1975=100)	1.80	0.48	0.54	0.23	0.06	0.08	0.07	0.05	0.06	0.03
GDP price deflator (1965=1.00)	0.10	0.07	–	–	–	–	–	–	–	–
Total employment	7.30	–0.03	2.60	1.20	1.30	0.61	0.28	0.22	0.63	0.50
Copper production (tonnes)	18.5	4.50	3.20	2.2	1.7	1.4	1.1	0.87	0.63	0.50
Exports of goods and services	81.40	60.80	3.00	2.00	1.20	0.94	1.10	1.10	0.45	0.46
Imports of goods and services	39.40	8.90	7.10	8.00	7.80	0.70	0.44	0.76	0.51	0.43
Balance of trade	50.70	53.90	–2.40	–4.20	–5.00	0.39	0.85	0.45	0.06	0.08
Balance of current account	15.60	53.60	–9.90	–11.00	–10.40	–2.10	–1.40	–1.70	–1.80	–1.60

Table 5-3
Percentage Differences between Shocked and Base Solutions for a 10 Percent Increase in Copper Prices in 1968

Variables	*1968*	*1969*	*1970*	*1971*	*1972*	*1973*
Real GDP	0.55	0.67	0.38	0.19	0.12	0.10
Agriculture	0.00	0.01	0.14	0.14	0.10	0.09
Mining and quarrying	1.32	0.34	0.29	0.21	0.16	0.14
Manufacturing	0.00	0.24	0.70	0.58	0.50	0.44
Services	–0.27	0.98	–0.04	–0.25	–0.16	–0.09
Copper production (tonnes)	0.65	0.44	0.32	0.26	0.20	0.16
Private consumption, real	2.52	0.81	–0.23	–0.77	–0.73	–0.60
Public consumption, real	–0.74	2.28	0.02	–0.37	–0.24	–0.13
Real gross investment	2.31	6.27	2.83	0.46	0.32	0.25
Total urban employment	–0.01	0.78	0.36	0.36	0.17	0.08
Employment in copper mining	0.00	2.00	1.30	2.50	1.70	1.10
Total wage bill	0.82	2.55	0.53	–0.28	–0.18	–0.09
Total operating surplus	15.42	0.41	0.73	0.38	0.19	0.15
Copper-mining profits	18.16	–0.26	0.20	0.33	0.22	0.17
Money supply	5.76	4.41	2.81	1.85	1.50	0.98
GDP deflator	5.17	0.53	0.24	–0.02	0.03	0.04
Consumer price index	0.76	0.78	0.32	0.08	0.10	0.08
Government current expenditure	1.03	6.90	0.85	–0.55	–0.55	–0.09
Government direct investment	2.54	7.82	0.63	–0.74	–0.42	–0.11
Government current revenues	14.81	1.72	–0.97	–0.66	–0.25	0.05
Imports of goods and services	1.93	1.64	1.73	1.39	0.13	0.08
Exports of goods and services	10.74	0.36	0.32	0.23	0.17	0.14
Net international reserves	27.20	12.83	8.61	12.00	20.17	10.42

increase in prices was such that after three years there was a negative impact on final demand. This inflationary effect is due mainly to the persistence of increases in the money supply brought about the increased inflow of foreign exchange.

Government current revenue initially increases then falls after two years. The explanation for the fall in government revenue is the decline of revenue from mining. Increases in prices lead to higher profits in the year of the increase and higher investments and labor costs in subsequent years. Profits in the subsequent years do not change very much, hence higher investments lead to a decline in the mineral-tax base and therefore a decline in mineral revenues. The decline in government revenues lead to declines in current and capital expenditures. However, the overall effect of the shock is to increase government revenues and expenditures.

A 10 Percent Decrease in the LME Price of Copper in 1968. Table 5–4 shows multiplier calculations and differences between the values of shocked and base solutions and table 5–5 presents percentage differences for the two solutions for major variables. The tables show that the decrease has a contractionary effect on

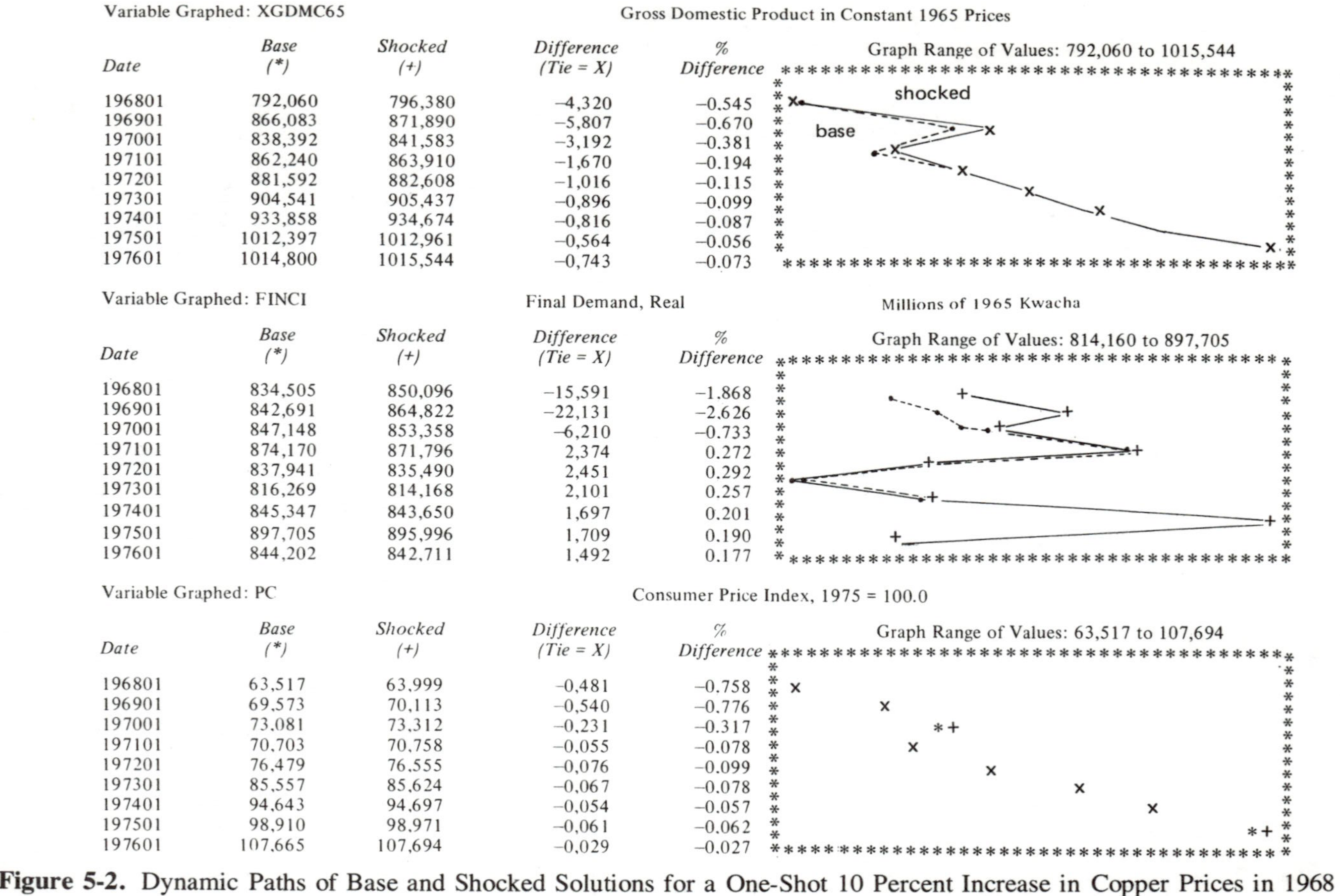

Variable Graphed: XGDMC65 — Gross Domestic Product in Constant 1965 Prices

Date	Base (*)	Shocked (+)	Difference (Tie = X)	% Difference
196801	792,060	796,380	–4,320	–0.545
196901	866,083	871,890	–5,807	–0.670
197001	838,392	841,583	–3,192	–0.381
197101	862,240	863,910	–1,670	–0.194
197201	881,592	882,608	–1,016	–0.115
197301	904,541	905,437	–0,896	–0.099
197401	933,858	934,674	–0,816	–0.087
197501	1012,397	1012,961	–0,564	–0.056
197601	1014,800	1015,544	–0,743	–0.073

Variable Graphed: FINCI — Final Demand, Real

Date	Base (*)	Shocked (+)	Difference (Tie = X)	% Difference
196801	834,505	850,096	–15,591	–1.868
196901	842,691	864,822	–22,131	–2.626
197001	847,148	853,358	–6,210	–0.733
197101	874,170	871,796	2,374	0.272
197201	837,941	835,490	2,451	0.292
197301	816,269	814,168	2,101	0.257
197401	845,347	843,650	1,697	0.201
197501	897,705	895,996	1,709	0.190
197601	844,202	842,711	1,492	0.177

Variable Graphed: PC — Consumer Price Index, 1975 = 100.0

Date	Base (*)	Shocked (+)	Difference (Tie = X)	% Difference
196801	63,517	63,999	–0,481	–0.758
196901	69,573	70,113	–0,540	–0.776
197001	73,081	73,312	–0,231	–0.317
197101	70,703	70,758	–0,055	–0.078
197201	76,479	76,555	–0,076	–0.099
197301	85,557	85,624	–0,067	–0.078
197401	94,643	94,697	–0,054	–0.057
197501	98,910	98,971	–0,061	–0.062
197601	107,665	107,694	–0,029	–0.027

Figure 5-2. Dynamic Paths of Base and Shocked Solutions for a One-Shot 10 Percent Increase in Copper Prices in 1968

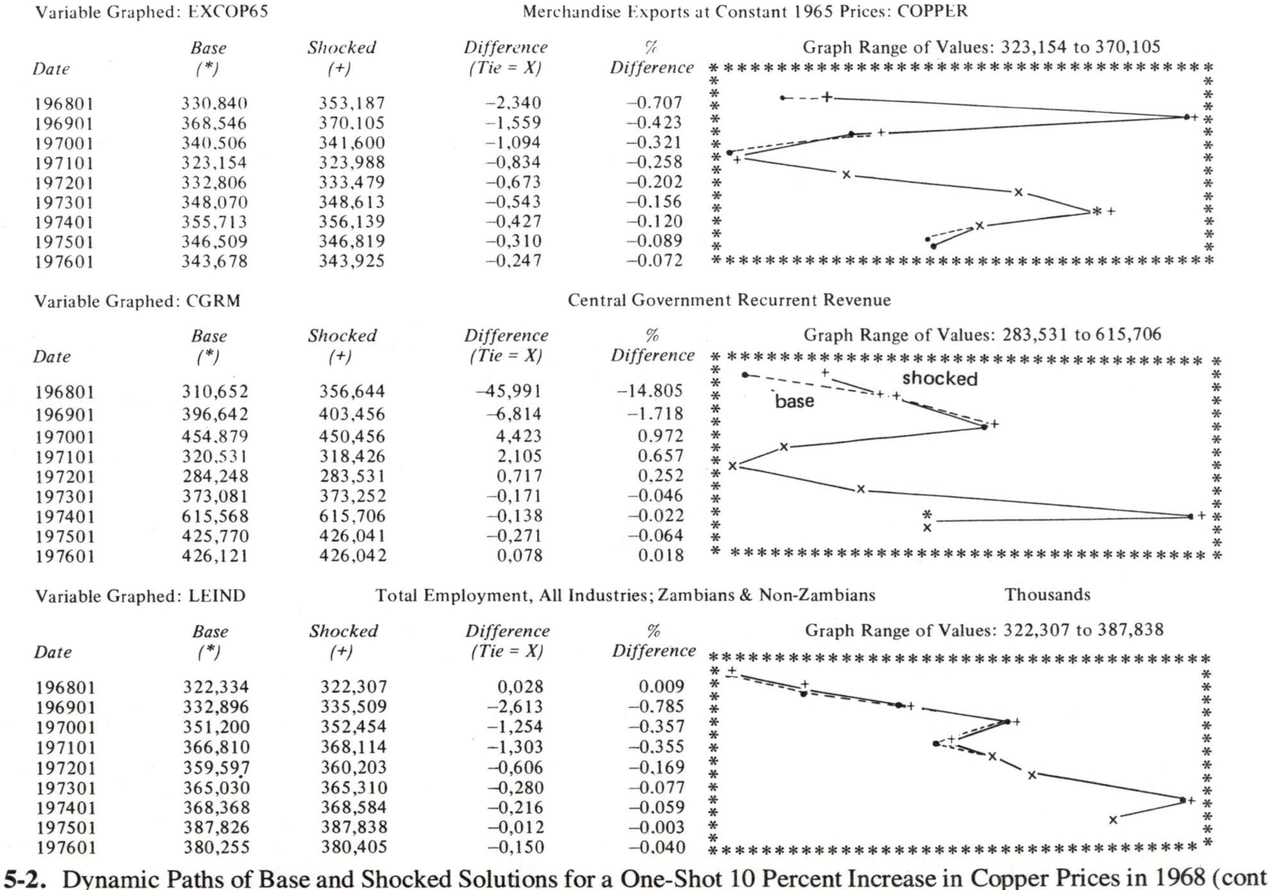

Variable Graphed: EXCOP65 — Merchandise Exports at Constant 1965 Prices: COPPER

Date	Base (*)	Shocked (+)	Difference (Tie = X)	% Difference
196801	330.840	353,187	−2,340	−0.707
196901	368.546	370.105	−1,559	−0.423
197001	340.506	341,600	−1,094	−0.321
197101	323.154	323,988	−0,834	−0,258
197201	332,806	333,479	−0,673	−0.202
197301	348,070	348,613	−0,543	−0.156
197401	355,713	356,139	−0,427	−0.120
197501	346,509	346,819	−0,310	−0.089
197601	343,678	343,925	−0,247	−0.072

Variable Graphed: CGRM — Central Government Recurrent Revenue

Date	Base (*)	Shocked (+)	Difference (Tie = X)	% Difference
196801	310,652	356,644	−45,991	−14.805
196901	396,642	403,456	−6,814	−1.718
197001	454.879	450,456	4,423	0.972
197101	320.531	318,426	2,105	0.657
197201	284,248	283,531	0,717	0,252
197301	373,081	373,252	−0,171	−0.046
197401	615,568	615,706	−0,138	−0.022
197501	425,770	426,041	−0,271	−0.064
197601	426,121	426,042	0,078	0.018

Variable Graphed: LEIND — Total Employment, All Industries; Zambians & Non-Zambians — Thousands

Date	Base (*)	Shocked (+)	Difference (Tie = X)	% Difference
196801	322,334	322,307	0,028	0.009
196901	332,896	335,509	−2,613	−0.785
197001	351,200	352,454	−1,254	−0.357
197101	366,810	368,114	−1,303	−0.355
197201	359,597	360,203	−0,606	−0.169
197301	365,030	365,310	−0,280	−0.077
197401	368,368	368,584	−0,216	−0.059
197501	387,826	387,838	−0,012	−0.003
197601	380,255	380,405	−0,150	−0.040

Figure 5-2. Dynamic Paths of Base and Shocked Solutions for a One-Shot 10 Percent Increase in Copper Prices in 1968 (cont.)

Table 5-4
Multipliers and Differences between Shocked and Base Solutions for a 10 Percent Decrease in Copper Prices in 1968

Variables	Dynamic Multipliers	Differences in Solutions' Values								
		1968	*1969*	*1970*	*1971*	*1972*	*1973*	*1974*	*1975*	*1976*
GDP	113.96	–62.90	–17.10	–8.50	–2.00	–1.60	–2.10	–2.10	–1.70	–1.40
GDP, real	20.80	–4.70	–5.90	–3.40	–1.80	–1.00	–0.92	–0.84	–0.55	–0.74
Government-consumption expenditure	6.20	–0.05	–6.00	–0.81	0.75	0.41	0.18	0.06	–0.02	0.04
Government current expenditure	21.10	–2.10	–16.80	–2.60	1.80	0.93	0.34	–	–0.03	–0.01
Government capital expenditure	17.00	–3.20	–15.10	1.10	2.00	0.46	0.08	–0.11	–0.04	–0.06
Government current revenue	52.20	–45.60	–7.00	4.30	2.20	0.80	–0.12	–0.09	–0.22	0.11
Final demand, real	38.20	–15.50	–22.20	–6.40	2.10	2.20	1.90	1.50	1.60	1.40
Gross fixed investment, real	36.40	–6.00	–15.00	–7.30	–1.30	–0.75	–0.51	–0.45	–0.23	–0.21
Money supply	69.00	–13.00	–13.40	–9.70	–5.40	–4.70	–4.30	–3.80	–3.20	–2.70
Consumer price index (1975=100)	1.80	–0.51	–0.56	–0.23	–0.03	–0.06	–0.05	–0.04	–0.05	–0.02
GDP price deflator (1965=1.00)	0.10	–0.07	–	–	–	–	–	–	–	–
Total employment	7.50	0.03	–2.60	–1.30	–1.40	–0.62	–0.30	–0.23	–0.02	–0.15
Copper production (tonnes)	20.30	–4.90	–3.50	–2.50	–1.90	–1.50	–1.20	–0.94	–0.68	–0.54
Exports of goods and services	82.10	–60.30	–3.30	–2.20	–1.30	–1.00	–1.30	–1.10	–0.49	–0.49
Imports of goods and services	42.50	–8.80	–7.10	–9.90	–8.80	–0.57	–0.39	–0.72	–0.43	–0.35
Balance of trade	49.00	–53.50	2.00	5.40	5.70	–0.55	–0.97	–0.55	–0.11	–0.15
Balance of current account	14.40	–53.20	9.50	12.80	11.20	1.70	1.10	1.40	1.60	1.40

Table 5-5
Percentage Differences between Shocked and Base Solutions for a 10 Percent Decrease in Copper Prices in 1968

Variables	*1968*	*1969*	*1970*	*1971*	*1972*	*1973*
Real GDP	–0.59	–0.68	–0.40	–0.21	–0.12	–0.10
Agriculture	0.00	–0.01	–0.14	–0.14	–0.10	–0.09
Mining and quarrying	–1.51	–0.38	–0.32	–0.32	–0.23	–0.15
Manufacturing	0.00	–0.24	–0.70	–0.58	–0.49	–0.43
Services	0.29	–0.98	0.03	–0.24	–0.15	–0.08
Copper production (tonnes)	–0.71	–0.49	–0.36	–0.29	–0.22	–0.17
Private consumption, real	–2.53	–0.80	0.22	0.72	0.67	0.55
Public consumption, real	0.79	–2.27	–0.03	0.34	0.23	0.12
Real gross investment	–2.28	–6.31	–2.86	–0.47	–0.29	–0.23
Total urban employment	0.01	–0.79	–0.37	–0.38	–0.17	–0.08
Employment in copper mining	0.00	–2.00	–1.30	–2.60	–1.70	–1.10
Total wage bill	–0.86	–2.54	–0.56	0.29	0.19	0.11
Total operating surplus	–15.24	–0.46	–0.75	–0.37	–0.18	–0.16
Copper-mining profits	–17.95	0.21	–0.23	–0.32	–0.22	–0.18
Money supply	–5.73	–4.42	–2.64	–1.60	–1.31	–0.87
GDP deflator	–5.15	–0.54	–0.24	0.05	–0.01	–0.03
Consumer price index	–0.80	–0.81	–0.31	–0.04	–0.07	–0.06
Government current expenditure	–1.02	–6.84	–0.88	0.54	0.30	0.10
Government direct investment	–2.51	–7.76	–0.66	0.72	0.44	0.13
Government current revenues	–14.67	–1.76	0.94	0.69	0.28	–0.03
Imports of goods and services	–1.90	–1.64	–2.14	–1.57	–0.11	–0.07
Exports of goods and services	–10.66	–0.39	–0.35	–0.26	–0.18	–0.15
Net international reserves	–27.01	–12.84	–8.16	–10.64	–18.09	–9.46

the economy. The sizes of the multipliers are comparable to those of a 10 percent increase in LME price in 1968. The nominal multiplier of gross domestic product is lower, but the real multiplier is higher in the case of the decrease. This is mainly because real output of copper declined more for the price decrease than it increased for the price increase. On the demand side, the major effect is on gross fixed investment. Government revenues decline, and therefore government investments decline. The savings of the government also fall, leading to a reduction in its loans to and investments in the private sector. Copper-industry profits fall, resulting in reduced investment outlays by the copper sector. The decline in gross fixed investment leads to a fall in real output by the noncopper sectors, especially manufacturing and commercial agriculture.

A Sustained 10 Percent Increase in the LME Copper Price, 1968–1976. The effects of the increase in prices are shown by multiplier calculations for selected variables in table 5–6. As before, the multipliers are calculated as the change in the variable per 100 kwacha increase in the price of copper. Table 5–7 shows percentage differences between shocked and base solutions. Figure 5–3 shows the dynamic paths of the base and shocked solutions for some variables.

Table 5-6
Multipliers for a 10 Percent Sustained Increase in Copper Prices, 1968-1976

Variables	*1968*	*1969*	*1970*	*1971*	*1972*	*1973*	*1974*	*1975*	*1976*
GDP	72.60	107.10	106.80	109.30	108.10	103.50	109.50	130.20	113.60
GDP, real	4.95	12.90	19.50	27.20	26.70	18.20	18.30	32.10	30.00
GDP of noncopper sectors, real	1.33	7.14	12.70	15.80	14.20	9.75	10.60	19.80	15.30
Government-consumption expenditure	0.05	5.93	9.74	7.53	3.57	2.17	4.42	9.00	1.96
Government current expenditure	2.37	19.30	29.80	24.20	13.70	9.46	15.80	28.90	9.65
Government capital expenditure	3.68	19.40	24.00	14.70	7.78	7.40	13.40	20.50	3.39
Government current revenue	52.70	71.80	45.80	31.00	28.50	41.20	43.10	25.60	13.60
Final demand, real	17.90	45.20	60.10	58.10	43.90	31.50	34.90	53.50	34.80
Gross fixed investment, real	7.00	24.60	38.60	44.40	38.60	25.40	26.60	46.50	33.00
Money supply	15.00	28.10	40.70	55.60	58.40	46.40	47.50	69.00	53.00
Consumer price index (1975=100)	0.00	0.95	1.13	1.40	1.36	0.88	0.81	1.30	0.74
GDP price deflator (1965=1.0)	0.08	0.09	0.09	0.08	0.08	0.08	0.08	0.08	0.06
Total employment	–0.03	2.56	5.05	7.50	7.37	4.46	4.76	7.55	6.21
Copper production (tonnes)	5.16	7.65	9.83	14.60	17.30	12.90	12.20	20.30	16.70
Exports of goods and services	69.60	80.50	76.90	74.80	78.60	83.50	86.20	84.20	84.40
Imports of goods and services	10.20	20.10	37.30	51.60	49.00	38.20	46.70	90.10	66.80
Balance of trade	61.80	65.20	48.00	34.20	39.50	55.10	50.10	12.50	26.60
Balance of current account	61.40	56.80	28.30	3.32	9.32	32.20	24.90	–37.20	–4.01

Table 5-7
Differences between Shocked and Base Solutions for a 10 Percent Sustained Increase in Copper Prices, 1968-1976

Variables	*Percentage Differences*								
	1968	*1969*	*1970*	*1971*	*1972*	*1973*	*1974*	*1975*	*1976*
Real GDP	0.55	1.53	2.31	2.40	2.28	2.29	2.55	2.49	2.93
Agriculture	0.00	0.01	0.16	0.27	0.33	0.33	0.36	0.39	0.44
Mining and quarrying	1.32	2.02	2.93	4.39	4.64	4.41	4.59	4.07	6.34
Manufacturing	0.00	0.24	0.88	1.69	1.84	1.98	2.17	2.50	2.61
Services	–0.27	0.76	1.12	0.21	–0.31	–0.28	0.08	0.19	–0.33
Copper production (tonnes)	0.65	1.09	1.42	1.70	1.93	2.09	2.20	2.27	2.38
Private consumption, real	2.52	4.16	4.06	2.38	1.27	1.98	2.43	1.07	0.61
Public consumption, real	–0.74	1.66	2.54	0.62	–0.39	–0.41	0.49	0.58	–0.27
Real gross investment	2.31	10.65	15.13	11.84	11.31	13.32	14.61	13.33	14.94
Total urban employment	–0.01	0.79	1.43	1.55	1.54	1.39	1.68	1.53	1.62
Employment in copper mining	0.00	1.80	3.80	6.20	8.50	7.60	9.40	7.40	12.70
Total wage bill	0.82	3.26	4.22	3.02	1.95	1.98	2.47	2.67	0.70
Total operating surplus	15.42	17.81	17.02	13.16	13.94	17.48	18.69	14.18	30.08
Copper-mining profits	18.16	16.08	18.46	28.52	35.07	23.25	20.11	45.86	161.05
Money supply	5.76	9.54	11.01	12.47	12.21	10.77	11.32	9.23	8.14
GDP deflator (1965=1.00)	5.17	6.25	5.58	4.40	4.03	4.95	4.98	3.87	3.34
Consumer price index (1975=100.0)	0.76	1.40	1.54	1.50	1.34	1.18	1.11	1.03	0.68
Government current expenditure	1.03	8.13	10.09	5.61	3.37	3.18	5.00	4.37	1.69
Government direct investment	2.54	12.98	12.91	7.72	5.79	7.44	9.27	6.93	3.50
Government current revenues	14.81	18.67	10.02	7.36	7.54	12.57	9.11	4.72	3.17
Imports of goods and services	1.92	4.77	8.03	7.01	6.92	8.32	8.41	8.50	9.29
Exports of goods and services	10.74	9.75	11.97	11.27	10.60	11.46	11.41	10.90	11.12
	Differences								
Government net budget deficit (–)	40.72	34.16	–7.92	–5.98	5.28	27.70	18.21	–18.67	0.59
Copper-industry gross profits	55.53	82.54	64.25	46.45	48.71	81.18	92.06	46.21	65.96
Balance of trade	53.91	67.19	47.71	25.99	29.76	62.80	65.14	9.77	26.34
Balance of current account	53.59	58.55	28.17	2.53	7.02	36.61	32.33	–29.16	–3.98
Net international reserves	40.84	85.61	107.38	109.70	115.44	143.76	168.91	147.30	144.80

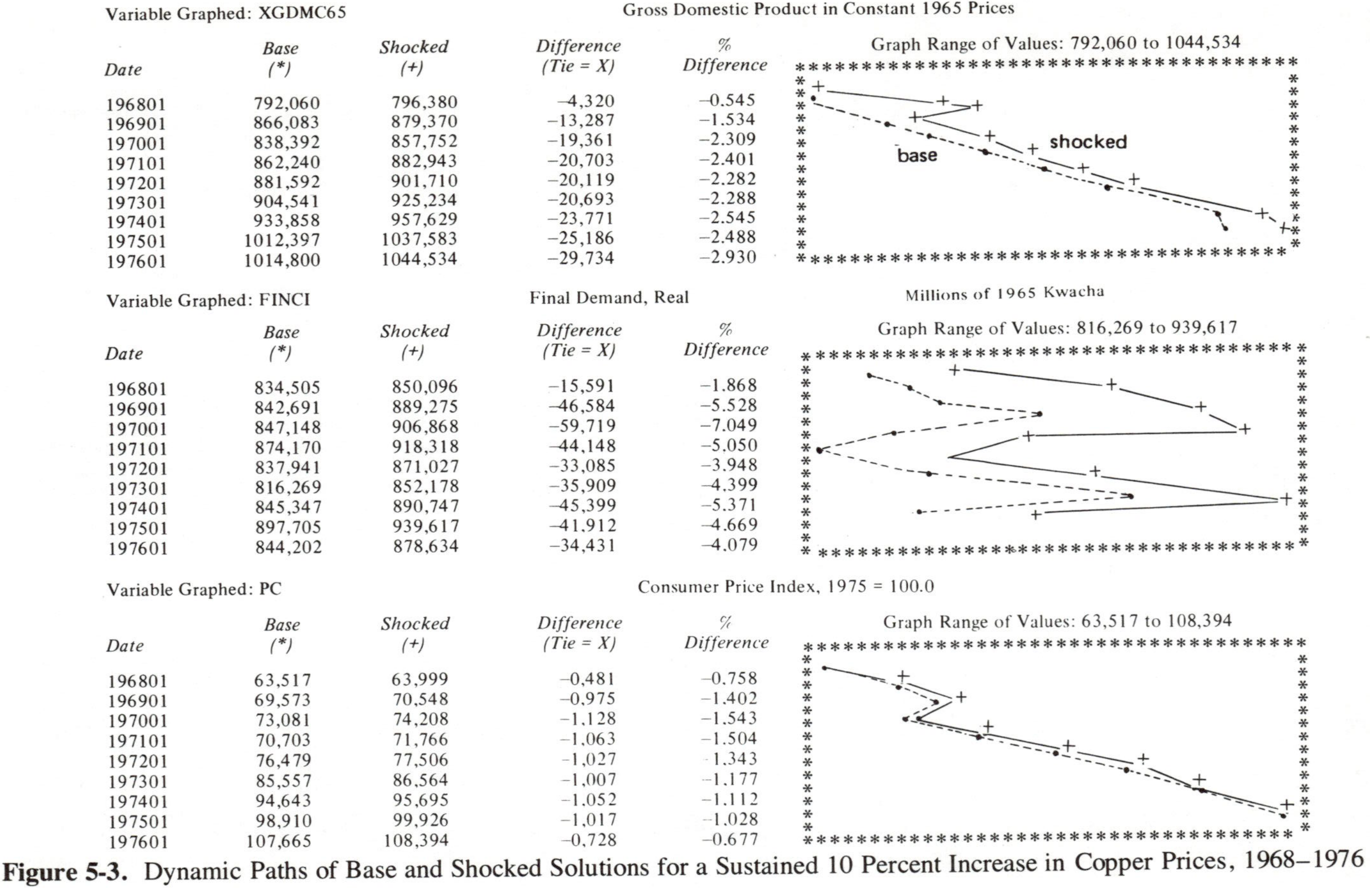

Variable Graphed: XGDMC65 — Gross Domestic Product in Constant 1965 Prices

Graph Range of Values: 792,060 to 1044,534

Date	Base (*)	Shocked (+)	Difference (Tie = X)	% Difference
196801	792,060	796,380	−4,320	−0.545
196901	866,083	879,370	−13,287	−1.534
197001	838,392	857,752	−19,361	−2.309
197101	862,240	882,943	−20,703	−2.401
197201	881,592	901,710	−20,119	−2.282
197301	904,541	925,234	−20,693	−2.288
197401	933,858	957,629	−23,771	−2.545
197501	1012,397	1037,583	−25,186	−2.488
197601	1014,800	1044,534	−29,734	−2.930

Variable Graphed: FINCI — Final Demand, Real — Millions of 1965 Kwacha

Graph Range of Values: 816,269 to 939,617

Date	Base (*)	Shocked (+)	Difference (Tie = X)	% Difference
196801	834,505	850,096	−15,591	−1.868
196901	842,691	889,275	−46,584	−5.528
197001	847,148	906,868	−59,719	−7.049
197101	874,170	918,318	−44,148	−5.050
197201	837,941	871,027	−33,085	−3.948
197301	816,269	852,178	−35,909	−4.399
197401	845,347	890,747	−45,399	−5.371
197501	897,705	939,617	−41,912	−4.669
197601	844,202	878,634	−34,431	−4.079

Variable Graphed: PC — Consumer Price Index, 1975 = 100.0

Graph Range of Values: 63,517 to 108,394

Date	Base (*)	Shocked (+)	Difference (Tie = X)	% Difference
196801	63,517	63,999	−0,481	−0.758
196901	69,573	70,548	−0,975	−1.402
197001	73,081	74,208	−1,128	−1.543
197101	70,703	71,766	−1,063	−1.504
197201	76,479	77,506	−1,027	−1.343
197301	85,557	86,564	−1,007	−1.177
197401	94,643	95,695	−1,052	−1.112
197501	98,910	99,926	−1,017	−1.028
197601	107,665	108,394	−0,728	−0.677

Figure 5-3. Dynamic Paths of Base and Shocked Solutions for a Sustained 10 Percent Increase in Copper Prices, 1968–1976

Variable Graphed: EXCOP65 — Merchandise Exports at Constant 1965 Prices: COPPER

Graph Range of Values: 323,154 to 372,540

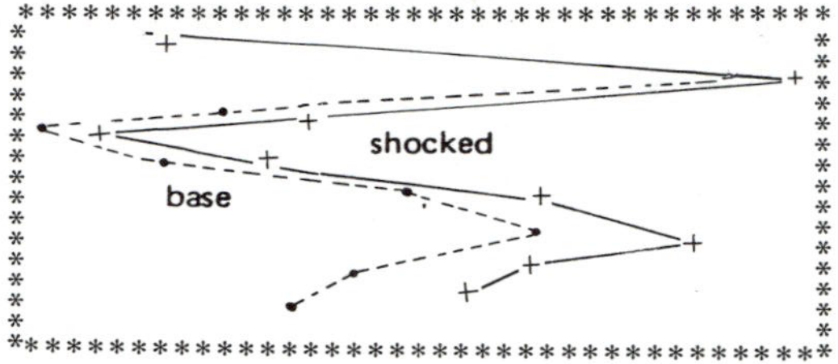

Date	Base (*)	Shocked (+)	Difference (Tie = X)	% Difference
196801	330,846	333,187	−2,340	−0.707
196901	368,546	372,540	−3,994	−1.084
197001	340,506	345,417	−4,912	−1.443
197101	323,154	328,756	−5,603	−1.734
197201	332,806	339,349	−6,542	−1.966
197301	348,070	355,412	−7,343	−2.109
197401	355,713	363,596	−7,884	−2.216
197501	346,509	354,476	−7,967	−2.299
197601	343,678	351,922	−8,243	−2.399

shocked

base

Variable Graphed: CGRM — Central Government Recurrent Revenue

Graph Range of Values: 284,248 to 671,656

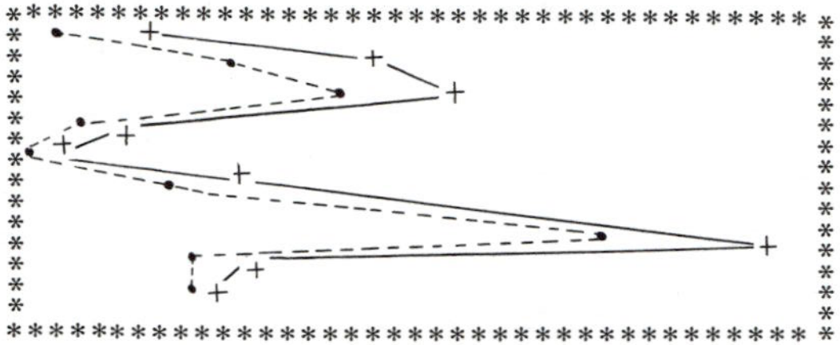

Date	Base (*)	Shocked (+)	Difference (Tie = X)	% Difference
196801	310,652	356,644	−45,991	−14.805
196901	396,642	470,693	−74,051	−18.669
197001	454,879	500,442	−45,563	−10.017
197101	320,531	344,111	−23,579	−7.356
197201	284,248	305,669	−21,421	−7.536
197301	373,081	419,983	−46,901	−12.571
197401	615,568	671,656	−56,088	−9.112
197501	425,770	445,844	−20,073	−4.715
197601	426,121	439,611	−13,490	−3.166

Variable Graphed: LEIND — Total Employment, All Industries: Zambians & Non-Zambians — Thousands

Graph Range of Values: 322,307 to 393,746

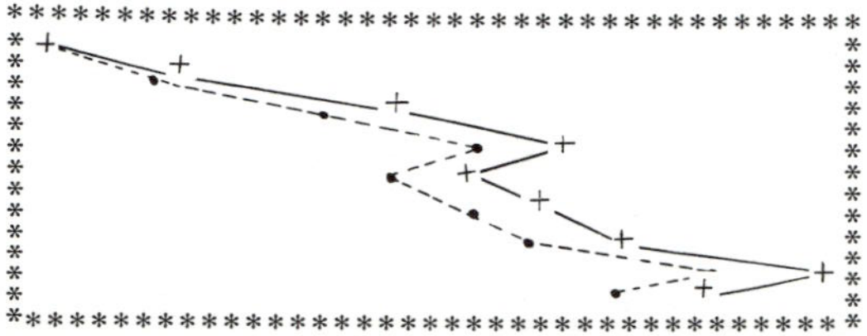

Date	Base (*)	Shocked (+)	Difference (Tie = X)	% Difference
196801	322,334	322,307	0,028	0.009
196901	332,896	335,536	−2,641	−0.793
197001	351,200	356,222	−5,022	−1.430
197101	366,810	372,509	−5,698	−1.554
197201	359,597	365,146	−5,550	−1.543
197301	365,030	370,107	−5,077	−1.391
197401	368,368	374,564	−6,196	−1.682
197501	387,826	393,746	−5,920	−1.527
197601	380,255	386,408	−6,153	−1.618

Figure 5-3. Dynamic Paths of Base and Shocked Solutions for a Sustained 10 Percent Increase in Copper Prices, 1968–1976 (cont.)

The sustained increase in prices has a sustained stimulative effect on the economy, with real gross domestic period increasing substantially throughout the period. There is increased real activity in the copper sector, which leads to greater exports. The most direct impacts are on balance of payments, international reserves, copper-industry profits, and investment. These, in turn, produce significant effects on government revenues and expenditures, public and private investment, money supply, and domestic prices. Real final demand shows a significant increase, which is dominated by increases in investment demand. In the simulation, the effect of large inflows of foreign assets on the money supply is not neutralized; this contributes to increases in the consumer price index. The contribution of the noncopper sectors to real gross domestic product increases, accounting for just over half the increases in real GDP in most years.

The multipliers for real GDP are highest in the years 1971, 1972, 1975, and 1976, when copper prices were low by historical standards. A combination of strong induced effects of past price increases and smaller levels of changes in these years (since the changes are on percentage basis) results in these higher multipliers. The year 1976 suffers from the dramatic fall of copper prices in 1975, but the multiplier is still high. The price increase seems to have had a perverse effect on the balance of current account in 1975 and 1976. The balance of trade increases, but, in 1975 and 1976, the effect of increased transfers abroad dominates the effect of increased trade balance. This is because the reserves accumulated in 1973 and 1974, when copper prices were high, make it possible to ease controls on the transfer of funds.

Table 5–8 shows the distribution of the induced effects of the increase in the prices of copper. The tabulated values are calculated as a percentage of the induced change in the nominal gross domestic product. Also shown, on the left-hand side of the table, is the average ratio of the variable and the nominal gross domestic product, calculated with the base-solution values. The table shows that the increase in prices has highly significant impacts on government revenues, operating surplus, and copper-industry profits. In these cases, the marginal ratios are much greater than the average ratios. The contribution of the copper sector to gross domestic product increases by more than its average values. Of the nonmining sectors, only construction and transportation and communications show marginal increases higher than average in the long run.

Sustained 10 Percent Decrease in LME Copper Prices, 1968-1976. Table 5–9 presents multiplier effects and table 5–10 shows the percentage differences from the base solution of a sustained 10 percent reduction in LME copper prices. The effect is contractionary, led by a decline in output of copper. This results in a fall in exports and worsening of the current account of the balance of payments. Imports also decline, since they depend on the level of domestic economic activity and the ability to import, measured by the amount of international reserves held. The decline in economic activity is more pronounced in those years in which copper prices were historically low—1970–1971 and 1975–1976. The effect of reduced

Table 5-8
Distribution of Induced Effects of a Sustained 10 Percent Increase in Copper Prices, 1968-1976

Variables	Average[a] Ratio	Induced Effect per Induced Effect on GDP								
		1968	*1969*	*1970*	*1971*	*1972*	*1973*	*1974*	*1975*	*1976*
Compensation of employees	42.30	5.29	14.50	21.20	17.90	13.10	10.40	12.70	20.60	6.20
Operating surplus	31.40	92.90	82.20	72.50	71.90	73.80	78.30	75.30	59.10	72.20
Capital consumption allowances	13.80	1.83	3.32	6.25	10.20	13.10	11.30	11.90	20.20	21.60
Copper-industry gross profits	17.00	87.70	74.70	60.60	56.00	59.80	68.90	64.60	45.30	58.60
Government mineral revenues	10.60	65.90	52.70	25.50	11.40	12.50	28.60	26.20	0.00	0.00
Investment income sent abroad	5.12	0.85	0.92	1.47	1.89	1.91	1.55	1.54	1.85	2.26
Private transfers abroad	5.13	0.34	7.16	15.20	24.30	25.50	18.60	19.00	31.00	24.50
Investment income from abroad	1.31	3.95	4.74	6.17	8.04	8.60	7.40	7.18	8.68	7.71
Contribution to GDP										
Mining and quarrying	28.80	88.50	78.00	65.90	60.90	64.00	72.00	69.50	52.10	62.80
Manufacturing	12.50	2.07	3.23	5.03	7.84	8.72	7.37	7.47	11.60	10.60
Construction	8.02	3.39	7.40	13.10	16.10	15.90	11.70	12.40	20.20	19.10
Transportation and communications	4.63	3.00	3.28	4.31	5.13	5.04	4.34	4.80	6.58	6.23
Services	33.40	2.92	7.77	11.10	9.54	6.11	4.44	5.61	9.06	1.59
Commercial agriculture	3.61	0.15	0.35	0.55	0.46	0.09	0.13	0.18	0.39	−0.42
Subsistence agriculture	7.99	−0.01	−0.02	−0.02	0.02	0.06	0.04	0.03	0.03	0.12

[a]Average ratio $= \frac{1}{9} \sum_{t=1}^{9} \frac{\text{Base-solution value of variable}}{\text{Base-solution GDP}}$

Table 5-9
Multipliers for a 10 Percent Sustained Decrease in Copper Prices, 1968-1976

Variables	*1968*	*1969*	*1970*	*1971*	*1972*	*1973*	*1974*	*1975*	*1976*
GDP	72.10	106.30	105.70	107.80	105.90	100.90	106.00	125.11	107.20
GDP, real	5.38	13.90	20.70	28.80	28.10	19.00	19.00	34.70	36.10
GDP of noncopper sectors, real	1.22	7.19	12.90	15.70	13.70	9.40	10.30	19.40	14.60
Government-consumption expenditure	0.06	5.88	9.64	7.36	3.33	1.96	4.14	8.39	1.57
Government current expenditure	2.36	19.20	29.50	23.70	13.00	8.87	15.00	27.00	8.61
Government capital expenditure	3.64	19.20	23.80	14.40	7.35	6.96	12.80	19.30	3.03
Government current revenue	52.20	71.10	45.10	29.70	26.60	39.40	40.70	23.90	11.20
Final demand, real	17.80	45.50	60.60	57.70	42.90	30.70	34.10	51.70	33.00
Gross fixed investment, real	6.90	24.90	39.20	44.80	38.50	25.20	26.40	45.80	32.30
Money supply	14.90	27.70	39.80	54.10	56.70	45.10	46.20	67.00	51.80
Consumer price index (1975=100)	0.58	1.00	1.22	1.51	1.47	0.93	0.82	1.31	0.67
GDP price deflator (1965=1.00)	0.08	0.10	0.09	0.08	0.08	0.08	0.07	0.07	0.04
Total employment	–0.03	2.55	5.01	7.60	7.44	4.42	4.82	7.33	6.45
Copper production (tonnes)	5.67	8.38	10.70	15.90	18.80	14.10	13.30	22.10	18.30
Exports of goods and services	69.10	79.60	75.80	73.40	77.00	81.70	84.30	82.30	82.50
Imports of goods and services	10.10	20.40	37.40	51.70	48.60	37.70	46.00	88.10	65.00
Balance of trade	61.40	64.00	46.70	32.80	38.20	53.70	48.70	12.02	26.20
Balance of current account	61.00	55.50	27.10	2.17	8.43	31.10	24.00	–36.40	–33.54

Table 5-10
Differences between Shocked and Base Solutions for a Sustained 10 Percent Reduction in LME Copper Prices

	Percentage Differences								
Variables	*1968*	*1969*	*1970*	*1971*	*1972*	*1973*	*1974*	*1975*	*1976*
Real GDP	−0.59	−1.65	−2.45	−2.54	−2.40	−2.39	−2.65	−2.69	−3.52
Agriculture	0.00	−0.01	−0.16	−0.27	−0.34	−0.34	−0.36	−0.39	−0.45
Mining and quarrying	−1.51	−2.35	−2.35	−3.37	−5.34	−5.01	−5.18	−5.07	−9.21
Manufacturing	0.00	−0.24	−0.88	−1.70	−1.84	−1.96	−2.14	−2.46	−2.56
Services	0.29	−0.74	−1.09	−0.12	0.42	0.37	0.00	−0.11	0.41
Copper production (tonnes)	−0.71	−1.19	−1.55	−1.85	−2.10	−2.27	−2.40	−2.47	−2.60
Private consumption, real	−2.53	−4.17	−4.06	−2.33	−1.17	−1.88	−2.29	−0.91	−0.36
Public consumption, real	0.79	−1.60	−2.46	−0.45	0.57	0.56	−0.38	−0.46	0.30
Real gross investment	−2.28	−10.80	−15.38	−11.92	−11.29	−13.21	−14.50	−13.13	−14.61
Total urban employment	0.00	−0.79	−1.44	−1.57	−1.55	−1.39	−1.70	−1.48	−1.68
Employment in copper mining	0.00	−1.80	−3.90	−6.50	−9.20	−8.20	−10.50	−8.00	−14.90
Total wage bill	−0.86	−3.29	−4.24	−3.02	−1.94	−1.91	−2.24	−2.43	−0.32
Total operating surplus	−15.24	−17.62	−16.75	−12.89	−13.57	−17.04	−18.26	−13.78	−29.38
Copper-industry gross profits	−18.00	−15.98	−18.09	−27.63	−33.62	−22.45	−19.33	−43.50	−150.06
Money supply	−5.73	−9.38	−10.77	−12.13	−11.86	−10.46	−11.02	−8.96	−7.95
GDP price deflator (1965=1.0)	−5.15	−6.28	−5.63	−4.39	−3.98	−4.90	−4.89	−3.62	−2.58
Consumer price index (1975=100)	−0.80	−1.49	−1.65	−1.63	−1.44	−1.24	−1.13	−1.04	−0.61
Government current expenditure	−1.02	−8.06	−9.99	−5.50	−3.21	−2.98	−4.74	−4.10	−1.51
Government direct investment	−2.51	−12.85	−12.76	−7.56	−5.53	−7.04	−8.82	−6.53	−3.17
Government current revenue	−14.67	−18.48	−9.85	−7.08	−7.05	−12.02	−8.59	−4.40	−2.60
Imports of goods and services	−1.90	−4.84	−8.06	−7.01	−6.87	−8.20	−8.29	−8.31	−9.04
Exports of goods and services	−10.66	−9.64	−11.80	−11.07	−10.39	−11.22	−11.16	−10.65	−10.86
	Differences								
Government net budget deficit (−)	−40.30	−33.80	8.10	6.30	−4.70	−26.80	−16.80	17.70	0.47
Copper-industry gross profits	−54.90	−81.50	−63.00	−45.00	−46.70	−78.40	−88.50	−43.80	−61.50
Balance of trade	−53.50	−66.00	−46.50	−25.00	−28.70	−61.10	−63.40	−9.40	−25.90
Balance of current account	−53.20	−57.20	−27.00	−1.70	−6.30	−35.40	−31.20	28.50	3.50
Net international reserves	−40.50	−84.30	−105.20	−106.80	−112.00	−139.40	−163.70	−142.60	−140.40

inflow of foreign exchange on the money supply dominates the effect of the increase in government domestic borrowing caused by a decline in government revenues, resulting in the fall of money supply. This reduction of the money supply is one of the factors contributing to a decline in domestic prices. The other factor is the reduction of wage rates caused by a drop in copper-company profitability and the fall of government revenues.

Effect of Changes in Copper Output

Copper output can increase exogenously, by opening a new mine, for example. It can also decrease exogenously by closure of a mine, caused by a willed decision, technical difficulties, or a major accident. It is assumed in the last two cases that no effort is made to step up output on other mines following the mine closure. Two types of changes will be examined: a sustained increase in output and a sustained decrease.

A Sustained 10 Percent Increase in Copper Output, 1968–1976. Copper output is endogenous in the model. For this test, it was exogenized at its base-solution values and increased by 10 percent in each year. Table 5–11 shows the percentage differences between base and shocked solutions. The increase in output results in increases of real gross domestic output, but with nearly all the increases coming from the copper sector. In some years, the gross domestic product of the noncopper sectors actually declines slightly. These are the years when lagged or contemporaneous copper prices are low. Therefore, the contributions of increased output to profits and hence to government revenues and wages are small. In 1975, profits of the industry decline with a 10 percent increase in output.

Increases in output of copper result in increased exports and imports, but the overall effect on foreign-exchange reserves is favorable. Increases in the capital account are less than those in the trade account because of increased outflows of investment incomes, service imports, and transfers. The rise in foreign assets leads to domestic monetary expansion, which is much greater than for the 10 percent sustained increase in copper prices. This is because, for output increase, the rise in government revenues is much smaller than for the price increase, and thus government budget surpluses, or decreases in deficit, do not offset the effect of increased foreign assets in the money-supply growth to the same extent. With a greater increase in foreign assets and a smaller rise in budget surplus, the effect of the increase in output is greater expansion of the money supply. The higher money supply leads to increases in the domestic price levels because of excess demand.

The shock results in a rise in total employment, although this is concentrated in the copper sector. Wages paid to Zambians and expatriates also rise because of increased productivity, which leads to higher wages for workers in other sectors, since their wages generally follow wages in the copper sector. The upward

Table 5-11
Differences between Shocked and Base Solutions for a Sustained 10 Percent Increase in Copper Output, 1968-1976

	Percentage Differences								
Variables	*1968*	*1969*	*1970*	*1971*	*1972*	*1973*	*1974*	*1975*	*1976*
GDP	3.92	5.08	5.58	4.71	4.46	5.27	5.31	3.56	4.12
GDP, real	1.76	2.72	2.68	1.72	1.51	2.06	2.12	0.82	2.75
GDP of noncopper sectors, real	–0.52	0.13	0.33	–0.14	–0.34	–0.23	0.25	–0.46	–0.21
Government-consumption expenditure	0.36	1.69	2.39	1.75	1.02	0.97	1.45	1.30	0.62
Government current expenditure	1.19	4.45	5.81	4.28	2.97	3.12	4.08	3.22	1.68
Government capital expenditure	0.84	6.96	7.68	5.79	3.83	5.60	7.08	4.75	2.67
Government current revenue	7.50	10.00	7.23	5.93	7.26	9.85	6.54	4.49	4.10
Final demand, real	0.76	2.62	3.00	1.41	0.41	1.05	1.57	0.19	–0.27
Gross fixed investment, real	–0.02	4.99	6.15	3.80	2.75	4.60	5.71	3.14	2.84
Money supply	7.20	11.20	12.80	15.60	16.20	14.20	15.00	14.40	14.00
Consumer price index (1975=100)	1.88	2.15	2.57	3.01	3.02	2.75	2.74	2.77	2.44
GDP price deflator (1965=1.0)	2.12	2.30	2.82	2.94	2.90	3.13	3.13	2.71	1.33
Total employment	0.24	1.14	1.41	1.62	1.49	1.58	1.67	1.67	1.38
Urban population	0.73	0.34	0.48	0.40	0.33	0.35	0.33	0.34	0.22
Copper production (tonnes)	10.00	10.00	10.00	10.00	10.00	10.00	10.00	10.00	10.00
Exports of goods and services	9.78	8.07	9.87	8.93	8.23	8.95	8.87	8.25	8.43
Imports of goods and services	0.69	1.81	3.26	2.79	2.15	3.31	3.82	2.55	2.57
	Differences								
Balance of trade	52.90	62.70	51.40	32.80	37.00	61.50	65.50	32.80	47.30
Balance of current account	53.60	56.60	37.20	14.30	18.10	39.20	37.70	2.86	19.20
Net international reserves	40.80	84.00	112.80	124.00	138.30	168.70	198.00	200.90	216.30
Copper-industry gross profits	23.60	40.20	30.70	11.40	12.50	32.70	35.20	–10.80	12.50

changes in the wage rates reinforce the increase in money supply, which causes domestic price levels to rise. Higher incomes result in higher public- and private-consumption expenditures in real terms for most years. Toward the end of the period, the effect of price level depresses real personal and public consumption until they fall below the base-solution values. Real gross fixed investment increases throughout the period, with investment in the copper sector dominating. Since real output in noncopper sectors was hardly affected by the shock, increases in final demand were met by increases in imports.

A Sustained 10 Percent Decrease in Copper Output, 1968–1976. Table 5–12 shows the effect of sustained exogenous decreases in copper production. The shock was applied by exogenizing copper production in the model and reducing the base-solution value by 10 percent. The effect of the shock is to decrease real output, employment, and the price level. The output and employment reductions are concentrated in the copper sector; the noncopper sector shows little change. The money supply contracts because of the fall in foreign assets; this effect dominates the increased need of the government to borrow from the Central Bank and commercial banks. Government revenues decline because of decreased mineral revenues, and expenditures by the government also decline as a response to decreased revenues. However, the overall effect of the shock is to increase the government budgetary deficit.

The reduction of output results in lower exports, copper-industry profits (except for 1975), and lower investment and other transfers abroad. Imports fall, partly because of a decreased domestic economic activity and partly because of a fall in ability to import. The decline in exports dominates the reduction in imports of goods and services; the net effects are declines of the balances of trade and current account and the net international reserves. Copper-industry profits increase in 1975, although gross profit is still negative. This is because the level of production in the base solution in 1975 is probably higher than its profit-maximizing level, given the cost structure of the industry.

Conclusions for Multiplier Tests

A few general observations can be made from these short-run multiplier tests and calculations of the impact of changes in the copper market on the Zambian economy. The most direct impact of changes in the copper market is on the balance of payments and international reserves, copper-industry profits, and investment. These produce significant induced effects on government revenues and expenditures, government and noncopper private-sector investment, and money supply and domestic prices.

Noncopper sectors of the economy are affected through induced changes in investment and government spending. These responses are generally lagged, since

Table 5-12
Differences between Shocked and Base Solutions for a Sustained 10 Percent Decrease in Copper Output, 1968-1976

Variables	1968	1969	1970	1971	1972	1973	1974	1975	1976
	Percentage Differences								
GDP	-4.00	-5.20	-5.70	-4.93	-4.70	-5.46	-5.50	-3.78	-4.34
GDP, real	-1.65	-2.66	-2.63	-1.56	-1.34	-1.96	-2.03	-0.71	-2.66
GDP of noncopper sectors, real	0.67	-0.04	-0.26	0.34	0.56	0.17	-0.14	0.19	0.13
Government-consumption expenditure	-0.41	-1.76	-2.48	-1.83	-1.11	-1.06	-1.53	-1.37	-0.68
Government current expenditure	-1.26	-4.60	-6.00	-4.44	-3.18	-3.34	-4.27	-3.38	-1.83
Government capital expenditure	-0.86	-7.09	-7.89	-5.99	-4.16	-5.95	-7.37	-4.97	-3.00
Government current revenue	-7.64	-10.30	-7.44	-6.33	-7.78	-10.30	-6.82	-4.89	-4.56
Final demand, real	0.73	-2.64	-3.05	-1.29	-0.23	-0.94	-1.49	-0.04	0.39
Gross fixed investment, real	0.27	-5.05	-6.40	-3.83	-2.73	-4.70	-5.87	-3.19	-2.89
Money supply	-7.26	-11.20	-12.80	-15.50	-16.10	-14.10	-14.90	-14.30	-13.90
Consumer price index (1970=100)	-2.08	-2.42	-2.87	-3.43	-3.47	-3.14	-3.11	-3.12	-2.75
GDP price deflator (1965=1.0)	-2.39	-2.61	-3.18	-3.42	-3.41	-3.58	-3.53	-3.09	-1.71
Total employment	-0.42	-1.12	-1.15	-1.66	-1.54	-1.65	-1.75	-1.75	-1.45
Urban population	-0.83	-0.43	-0.54	-0.47	-0.39	-0.41	-0.38	-0.39	-0.26
Copper production (tonnes)	-10.00	-10.00	-10.00	-10.00	-10.00	-10.00	-10.00	-10.00	-10.00
Exports of goods and services	-9.78	-8.07	-9.87	-8.93	-8.24	-8.95	-8.87	-8.25	-8.43
Imports of goods and services	0.59	-1.84	-3.34	-2.77	-2.07	-3.25	-3.84	-2.50	-2.54
	Differences								
Balance of trade	-53.30	-62.70	-51.10	-32.90	-37.20	-61.50	-65.40	-33.10	-47.40
Balance of current account	-54.00	-56.40	-36.70	-14.30	-18.30	-39.00	-37.20	-3.11	-19.30
Net international reserves	-41.40	-84.20	-112.50	-123.80	-138.20	-168.40	-197.50	-200.60	-216.00
Copper-industry gross profits	-22.40	-40.30	-30.50	-11.50	-12.40	-32.60	-35.30	10.80	-12.30

government spending responds to changes in revenues with a lag, and capital stock responds to investment changes with a lag. The impact of changes in the price of copper on the noncopper sectors is very weak, indicating that the linkages are not strong. There is little effect on agricultural output from changes in copper prices or activities, except for small effects caused by increments or decreases in capital stock on commercial agriculture and by the change in rural-urban migration on subsistence agriculture. These two induced changes act in offsetting directions and are small in the short run. For example, a sustained 10 percent increase in the price of copper results in an average percentage change of 0.15 percent in real agricultural output for the first five years. The effect of autonomous increases in copper production on noncopper sectors is even weaker than the price effect. The contribution to real gross domestic product by the noncopper sectors shows little change for a sustained 10 percent increase in copper output, declining by an average of −0.1 percent for the first five years. The effect of weak linkages between the copper sector and the rest of the economy is that the effect of fluctuations of prices on the noncopper sectors is reduced, although fluctuations may in themselves hamper the ability of the government to mobilize resources for development.

Changes in the commodity-sector activity have impact on the price level through the effects of changes in foreign-exchange earnings and government financing of deficit or surplus on the money supply. Movements in money supply caused by commodity-price or output changes dominate the impact of induced surplus or deficit of government spending on the money supply. In the simulations, government fiscal and monetary policies are assumed to be passive. The effect of increases in the price of copper, for example, is to raise the domestic price levels, primarily through changes in the money supply. The impact of the copper market on the domestic price levels calls for active monetary and fiscal policies to counteract these effects.

Decreases in the price of copper have a greater absolute effect than do increases in price on such variables as gross domestic product, investment, and total employment, although the differences are small. This tendency is obtained for both one-time changes and sustained changes in the prices of copper. The average percentage change of real gross domestic product for a sustained 10 percent increase in the price of copper in 1968–1972 was 1.81 percent, while the change of real gross domestic product for a sustained 10 percent decrease was −1.93 percent for the corresponding period. This result indicates that fluctuations of prices hinder economic growth. It implies that declines in copper prices cause a greater decline in economic activity, especially in the commodity sector, than the increase in activity caused by corresponding increases in copper prices.

Fluctuations in Copper Prices and Goal Attainment

We conclude this chapter by looking at the impact of copper-price fluctuations on the Zambian economy over a relatively long period of time. The procedure again involves generating a control solution that assumes a smooth trend in commodity prices and then comparing this to other solutions obtained from fluctuating prices. Since our interest is focused only on the impact of fluctuations, we induce random movements of prices along the smooth long-term trend, with a mean of zero and one or two standard deviations. This series extends the one-period analysis just discussed by taking into account the time dimension; it allows the lagged variables to make their impact on the system. Varying the degree of the fluctuations allows us to test for possible nonlinearities in the response pattern.

The time pattern of the random shocks may have an important bearing on the terminal values of the relevant impact variables. Thus, a concentration of downward shocks in the initial years of any single simulation run may easily outweigh the impact of upward shocks concentrated toward the end of the period, because the lags are such that they permit the full impact of the earlier movements to be felt but not those of the later movements. It is also possible to have different results from different runs, even if the mean copper-export price is the same, because the present values of the discounted prices may not be the same. A solution to this problem would be to conduct a large number of simulations and compare the results, taking into account the observed pattern of fluctuations.

The control solution for 1968–1990 involved a smooth upward trend in copper prices. A regression of the log of the historical copper price against a time trend gave an average annual rate of increase of 5.48 percent. This rate was then used to project prices to 1990. Other important exogenous variables involved imports and the total population. Fuel-import prices were assumed to rise at 6 percent and other import-price indexes by 5 percent per year. The total population was assumed to increase at the historical rate of 2.5 percent. Copper-ore grades were kept constant, as was the percentage distribution of output by method of mining. Table 5–13 gives the values of some of the key variables that were generated by the solution: copper production, total exports of goods and services, real final demand, gross private investment at both current and constant values, investment into mining, imports of goods and services, the urban population, employment and wages in copper mining, and total value added at current and constant prices.

Tables 5–14 through 5–18 give the same variables but from solutions obtained with fluctuating prices. The fluctuations were random, with a zero mean and one standard deviation from trend.

Several simulations were carried out, but the results were so consistent that only a few need be reported here. The five sets of results were based on randomly fluctuating price patterns that eventually covered any patterns that could be conceived. The first and fourth series (tables 5–14 and 5–17) started with relatively sharp downward shocks followed by fairly well dispersed and small upward movements. The fifth series (table 5–18) started with upward movements, followed by mild declines.

Again, the results show the macro variables to be relatively immune to price fluctuations. The terminal magnitudes of value added at constant prices differ by about 1.3 percent or less from the smooth-trend value. The simulation with a lower

Table 5-13
Control Solution: Smooth Copper Price Trend

	PCLME	*CUQCZM*	*WGAMQZ*	*XGDMC65*	*IGFC65*
1968	614.44	703.04	1244.20	793.45	275.04
1969	648.08	732.45	1634.00	853.93	269.87
1970	683.56	702.56	1570.40	841.54	294.62
1971	720.98	686.19	1211.70	888.30	349.09
1972	760.46	774.27	1518.10	975.37	384.29
1973	802.09	762.95	1403.90	937.47	361.50
1974	846.00	731.65	1373.00	913.56	485.93
1975	892.32	730.46	1385.20	955.68	310.35
1976	941.17	774.39	1916.50	962.36	260.99
1977	992.70	768.84	1709.00	1017.40	296.62
1978	1047.05	774.52	2055.10	1066.40	347.32
1979	1104.38	784.46	2139.00	1113.10	377.96
1980	1164.84	794.13	2292.30	1152.50	410.86
1981	1228.61	803.91	2359.00	1190.00	433.50
1982	1295.88	813.75	2434.80	1229.90	457.47
1983	1366.82	823.61	2515.50	1271.80	482.78
1984	1441.66	833.47	2595.60	1316.40	509.77
1985	1520.58	843.30	2679.70	1363.20	531.11
1986	1603.83	853.07	2768.70	1411.80	567.52
1987	1694.64	862.78	2860.30	1462.50	598.24
1988	1784.26	873.43	2954.70	1515.30	630.36
1989	1881.94	882.01	3051.80	1570.30	663.93
1990	1984.98	891.51	3158.20	1626.10	697.97

	LEIND	*XAFFS65*	*NINTR*	*PC*	*PGDP*
1968	324.94	81.83	239.69	64.63	1.5350
1969	328.86	81.51	367.86	71.46	1.8167
1970	355.73	81.96	479.78	75.16	1.7066
1971	372.15	82.77	540.57	74.80	1.7406
1972	377.51	82.90	718.68	83.33	1.9285
1973	386.81	83.88	699.09	89.07	1.9035
1974	387.07	84.53	628.80	97.18	1.9914
1975	392.59	86.19	726.95	104.29	1.9120
1976	386.37	87.12	752.24	120.71	2.0781
1977	407.37	87.69	812.87	118.00	2.2164
1978	416.13	88.02	798.22	120.91	2.3182
1979	427.74	88.52	854.54	123.99	2.4862
1980	440.07	88.99	877.54	126.83	2.5805
1981	451.92	89.53	892.52	129.14	2.6608
1982	463.61	90.08	904.89	131.29	2.7391
1983	475.49	90.65	915.15	133.35	2.8162
1984	487.74	91.25	930.92	135.14	2.8902
1985	500.31	91.88	952.02	136.90	2.9640
1986	513.13	92.54	975.66	138.75	3.3093
1987	526.28	93.23	1002.30	140.60	3.1154
1988	539.76	93.96	1032.30	142.46	3.1925
1989	553.55	94.73	1066.30	144.33	3.2707
1990	567.33	95.55	1091.60	146.52	3.3543

Table 5-14
Price Fluctuations: Series 1

	PCLME	*CUQCZM*	*WGAMQZ*	*XGDMC65*	*IGFC65*
1968	293.33	668.75	1118.70	744.35	235.40
1968	741.43	713.96	1442.80	838.35	202.44
1970	705.39	609.54	1720.80	840.34	294.26
1971	665.42	673.20	1334.80	889.66	352.89
1972	749.68	763.86	1484.40	970.61	374.91
1973	792.40	754.53	1402.40	932.50	347.79
1974	860.98	726.57	1369.90	917.12	282.91
1975	1051.57	734.96	1448.60	987.17	323.81
1976	953.54	777.72	2075.60	980.27	285.08
1977	982.77	770.67	1708.50	1026.30	320.87
1978	1183.49	782.38	2030.50	1086.50	360.23
1979	999.05	784.53	2197.40	1098.20	380.90
1980	1211.35	796.61	2179.90	1162.50	422.44
1981	1149.07	802.04	2358.80	1169.70	414.75
1982	1264.61	811.32	2350.70	1223.80	450.63
1983	1300.18	826.94	2597.60	1290.60	492.34
1985	1598.65	841.84	2572.70	1369.50	531.14
1986	1566.95	850.54	2859.90	1398.90	5490.05
1987	1836.60	865.80	2863.40	1486.10	620.14
1988	1568.99	866.79	3082.70	1482.70	622.12
1989	2010.16	882.43	2850.90	1591.00	678.04
1990	2124.62	895.50	3315.90	1630.90	684.93

	LEIND	*XAFFS65*	*NINTR*	*PC*	*PGDP*
1968	325.90	81.87	–38.32	60.67	1.1034
1969	312.19	81.60	248.12	68.75	1.3074
1970	354.75	81.85	383.50	74.63	1.4349
1971	368.52	82.68	389.21	74.04	1.4165
1972	376.24	82.90	570.25	81.99	1.5655
1973	385.57	83.87	563.84	88.03	1.6757
1974	385.92	84.51	519.30	96.34	1.7904
1975	392.98	86.16	738.70	104.26	1.9052
1976	390.16	87.04	732.56	120.87	2.2221
1977	409.99	87.69	749.28	117.45	2.1359
1978	418.29	88.04	837.81	120.83	2.2031
1979	431.03	88.49	786.97	123.94	2.2673
1980	439.99	89.05	844.77	126.13	2.3076
1981	452.80	89.53	821.46	128.93	2.3614
1982	461.91	90.12	825.18	130.64	2.3898
1983	473.91	90.67	886.89	133.17	2.4372
1984	486.31	91.25	798.21	134.79	2.4646
1985	496.90	91.93	926.28	136.33	2.4909
1986	512.12	92.49	951.18	139.13	2.5449
1987	524.93	93.23	1086.60	140.84	2.5727
1988	541.31	93.89	892.96	142.75	2.6055
1989	551.03	94.84	1045.40	143.33	2.6071
1990	567.97	95.47	1251.70	147.61	2.6924

Table 5-15
Price Fluctuations: Series 2

	PCLME	*CUQCZM*	*WGAMQZ*	*XGDMC65*	*IGFC65*
1968	704.99	709.75	1270.90	801.17	285.87
1969	475.57	721.93	1642.70	837.07	264.78
1970	588.56	688.79	1398.60	809.65	228.77
1971	602.02	668.83	1156.50	855.26	286.95
1972	978.44	772.89	1632.90	985.38	272.38
1973	564.08	741.44	1281.60	917.64	264.25
1975	755.14	711.49	1429.60	913.61	271.72
1976	905.79	758.63	1970.40	978.05	242.99
1977	963.26	755.98	1764.00	1008.60	274.28
1978	965.57	761.02	2108.80	1058.60	331.07
1979	1097.73	774.53	2145.60	1113.00	365.49
1980	1001.22	779.03	2316.20	1124.30	382.40
1981	1069.09	786.14	2250.00	1161.10	395.05
1982	1163.06	795.71	2338.10	1196.20	400.53
1983	1240.36	805.84	2468.00	1242.20	429.79
1984	1549.61	825.10	2600.30	1323.60	484.42
1985	1595.10	839.76	2892.30	1374.50	537.79
1986	1570.12	849.20	2907.90	1421.70	593.37
1987	1568.65	859.07	2859.50	1465.10	607.19
1988	1607.07	863.97	2922.10	1487.90	606.61
1989	1861.63	875.94	2903.20	1562.50	638.28
1990	1878.67	884.17	3172.10	1600.10	655.82

	LEIND	*XAFFS65*	*NINTR*	*PC*	*PGDP*
1968	324.81	81.82	315.96	65.51	1.6620
1969	333.51	81.49	262.23	70.88	1.5594
1970	346.99	82.05	337.77	73.36	1.5699
1971	365.24	82.79	349.49	72.99	1.5898
1972	366.59	82.83	952.56	83.79	2.2111
1973	391.29	83.71	528.08	89.03	1.7098
1974	380.70	84.57	501.08	95.66	1.9784
1975	391.80	86.16	542.94	103.23	1.8154
1976	380.30	87.08	576.85	119.48	2.0162
1977	405.76	87.65	656.08	117.07	2.1810
1978	413.93	87.98	606.95	119.83	2.2503
1979	425.83	88.50	687.97	122.89	2.4630
1980	438.78	88.96	613.31	125.54	2.4991
1981	447.89	89.57	558.55	127.02	2.5149
1982	458.19	90.14	573.36	129.34	2.6124
1983	469.14	90.66	585.06	131.69	2.7045
1984	481.64	91.24	779.87	134.33	2.9511
1985	498.65	91.76	878.26	137.35	3.0338
1986	513.72	92.46	808.79	138.66	3.0305
1987	526.80	93.22	792.60	139.80	3.0846
1988	539.45	93.96	709.88	141.24	3.0696
1989	550.33	94.80	814.32	142.75	3.2268
1990	564.69	95.53	851.61	145.79	3.2849

Table 5-16
Price Fluctuations: Series 3

	PCLME	*CUQCZM*	*WGAMQZ*	*XGDMC65*	*IGFC65*
1968	421.30	685.36	1176.40	771.25	252.70
1969	541.23	711.16	1457.40	826.86	199.64
1970	956.45	703.77	1601.70	850.35	275.08
1971	616.94	678.04	1506.10	893.50	392.34
1972	554.23	753.02	1406.30	951.95	368.42
1973	1002.02	760.38	1195.80	937.54	320.41
1974	785.76	725.23	1579.20	909.33	292.51
1975	903.89	726.47	1381.90	968.50	326.99
1976	943.22	771.69	1912.60	957.30	255.72
1977	785.57	754.81	1678.90	969.20	276.52
1978	1281.76	775.46	1949.00	1090.70	344.28
1979	1159.57	787.14	2355.90	1118.00	378.32
1980	1324.03	802.68	2389.90	1191.50	468.77
1981	1219.02	809.24	2466.70	1195.10	462.76
1982	1176.74	812.41	2356.60	1215.90	469.35
1983	1415.89	825.00	2367.60	1270.80	471.15
1984	1423.65	833.68	2619.00	1305.40	492.66
1985	1600.34	846.36	2670.50	1375.60	550.48
1986	1573.95	853.96	2835.70	1406.60	570.95
1987	1683.23	803.21	2818.10	1463.90	607.42
1988	1929.89	877.28	2953.20	1530.20	636.97
1990	2148.12	896.63	3075.10	1653.60	729.35

	LEIND	*XAFFS65*	*NINTR*	*PC*	*PGDP*
1968	325.45	81.85	67.31	62.45	1.2686
1969	319.26	81.58	144.63	68.37	1.6209
1970	345.40	81.94	553.69	75.28	2.0353
1971	376.72	82.58	467.33	75.67	1.6577
1972	375.41	82.95	432.01	81.08	1.6566
1973	380.10	84.00	668.23	87.73	2.0686
1974	388.70	84.41	534.32	97.31	1.9522
1975	390.78	86.19	625.71	103.47	1.9074
1976	386.93	87.12	668.29	120.12	2.0662
1977	406.15	87.70	595.76	116.74	2.0786
1978	411.94	88.08	803.29	120.26	2.4616
1979	429.20	88.41	905.95	125.16	2.5527
1980	441.63	88.95	975.74	127.51	2.7178
1981	457.01	89.47	919.11	129.74	2.6689
1982	466.00	90.12	793.68	130.61	2.6483
1983	474.92	90.73	891.32	132.67	2.8263
1984	487.62	91.24	927.44	135.34	2.8804
1985	499.59	91.89	1006.30	137.10	3.0182
1986	514.05	92.50	986.88	139.20	3.0307
1987	526.09	92.25	985.82	140.47	3.1059
1988	539.65	93.96	1157.60	142.85	3.2854
1989	555.49	96.66	1075.70	145.07	3.2433
1990	567.49	95.60	1200.70	146.48	3.4434

Table 5-17
Price Fluctuations: Series 4

	PCLME	*CUQCZM*	*WGAMQZ*	*XGDMC65*	*ICFC65*
1968	487.68	692.17	1201.80	780.18	260.23
1969	642.20	724.30	1593.30	843.51	235.39
1970	640.73	693.77	1581.30	828.84	272.41
1971	810.28	685.16	1270.90	895.35	346.38
1972	713.42	769.69	1630.70	975.40	393.42
1973	843.33	762.20	1388.20	945.09	365.94
1974	748.02	724.76	1379.00	895.62	280.16
1975	839.08	722.69	1294.80	941.24	294.13
1976	899.79	766.73	1876.30	943.71	237.08
1977	1062.20	766.92	1728.60	1026.70	289.61
1978	952.75	768.01	2129.60	1050.90	341.20
1979	906.77	769.72	2056.30	1077.00	358.66
1980	1051.64	778.75	2123.60	1119.30	360.39
1981	1158.78	790.07	2305.70	1164.00	381.05
1982	1164.36	798.25	2437.20	1201.50	418.97
1983	1258.68	808.41	2466.30	1248.00	447.19
1984	1444.51	823.04	2581.40	1307.30	477.08
1985	1563.86	837.47	2775.50	1365.70	525.55
1986	1593.89	848.51	2882.60	1415.30	573.93
1987	1711.20	860.28	2902.50	1470.40	608.57
1988	1725.09	868.73	2999.20	1508.70	627.32
1989	1879.54	879.53	3018.30	1571.40	662.13
1990	2023.89	890.96	3187.10	1628.10	691.56

	LEIND	*XAFFS65*	*NINTR*	*PC*	*PGDP*
1968	325.26	81.84	124.68	63.25	1.3589
1969	322.50	81.55	280.65	70.08	1.7921
1970	352.79	81.94	372.67	74.24	1.6439
1971	367.51	82.73	529.21	74.74	1.8432
1972	379.27	82.83	644.98	83.29	1.8774
1973	386.72	83.88	666.48	88.72	1.9387
1974	388.64	84.52	517.82	96.53	1.9024
1975	390.16	86.24	603.98	103.13	1.8545
1976	383.95	87.14	643.29	119.93	2.0398
1977	404.66	87.98	679.10	120.74	2.2611
1979	425.31	88.55	608.17	122.52	2.3255
1980	434.61	89.07	632.47	125.07	2.4619
1981	445.63	89.55	692.91	128.10	2.5900
1982	457.76	90.07	661.30	130.35	2.6374
1983	469.24	90.67	650.24	132.13	2.7241
1984	481.83	91.25	751.22	134.37	2.8792
1985	496.55	91.82	849.52	136.90	2.9994
1986	511.68	92.47	849.93	138.80	3.0442
1987	525.58	93.20	879.92	140.36	3.1290
1988	539.78	93.93	862.30	142.14	3.1570
1989	552.65	94.74	912.50	143.70	3.2597
1990	566.84	95.53	1007.50	146.37	3.3777
1989	550.33	94.80	814.32	142.75	3.2268
1990	564.69	95.53	851.61	145.79	3.2849

Table 5-18
Price Fluctuations: Series 5

	PCLME	*CUQCZM*	*WGAMQZ*	*XGDMC65*	*IGFC65*
1968	515.04	707.37	1261.80	798.57	282.03
1969	538.43	727.51	1646.10	846.32	269.03
1970	605.02	709.05	1523.70	851.57	288.77
1971	746.61	693.43	1339.00	903.40	386.11
1972	792.54	788.61	1585.10	1002.40	436.78
1973	734.13	776.87	1488.00	959.94	416.46
1974	857.92	749.59	1337.10	940.25	329.17
1975	1047.59	739.65	1352.60	939.65	336.65
1976	854.65	776.51	1694.30	904.64	253.72
1977	937.97	773.83	1572.70	1017.60	280.34
1978	1071.86	783.94	2100.20	1082.10	350.75
1979	1023.17	783.96	2202.90	1095.30	388.94
1980	996.85	794.47	2146.30	1154.20	414.70
1981	1150.52	799.64	2323.40	1163.10	403.72
1982	1377.60	811.31	2347.60	1228.40	446.27
1983	1359.28	821.13	2531.00	1263.60	466.55
1984	1485.28	835.38	2607.10	1331.90	517.77
1985	1543.19	839.19	2761.40	1344.60	535.01
1986	1476.79	847.15	2625.70	1400.70	560.83
1987	1712.94	855.44	2760.90	1436.00	556.35
1988	1857.29	864.15	2866.20	1429.50	593.66
1989	1858.72	876.47	2987.20	1562.20	635.97
1990	2142.62	888.34	3203.80	1623.90	681.94

	LEIND	*XAFFS65*	*NINTR*	*PC*	*PGDP*
1968	324.86	81.82	287.60	65.21	1.6172
1969	331.86	81.50	313.80	71.23	1.6725
1970	351.23	82.00	584.09	75.59	1.8909
1971	377.30	82.70	657.59	76.22	1.8220
1972	381.06	82.88	953.31	85.06	2.1374
1973	396.36	83.84	923.29	90.80	1.9940
1974	394.29	84.56	902.40	98.71	2.1236
1975	400.42	86.22	890.45	105.36	1.9001
1976	389.21	87.24	846.28	120.75	2.0759
1977	407.60	87.77	976.76	118.50	2.2549
1978	416.62	88.01	1050.40	122.46	2.4293
1979	429.96	88.49	936.38	124.85	2.4156
1980	439.20	89.07	954.12	126.61	2.5758
1981	451.56	89.55	919.21	129.25	2.5925
1982	461.09	90.13	958.34	131.13	2.7386
1983	474.33	90.65	975.05	133.65	2.8092
1984	486.71	91.25	1065.00	135.57	2.9634
1985	501.56	91.84	939.27	137.23	2.8927
1986	511.61	92.61	883.28	137.77	2.9599
1987	523.77	93.28	908.75	139.80	3.0352
1900	536.04	94.00	922.76	141.65	3.1149
1989	549.52	94.76	1029.70	143.80	3.2593
1990	564.76	95.52	1157.70	146.68	3.3748

terminal value of the LME price (table 5–15) also has a slightly lower terminal value added. The impact is largest on the net international reserves and the value of total exports of goods and services. The elasticity of these variables with respect to a price shock is positive and greater than unity. Again, the effects are not passed on to the rest of the economy. Employment of both types of labor responds to the long-term trend rather than to short-run cycles, as do mining wages and real output from the agricultural sector.

The same response pattern was observed when we employed a smooth downard trend to LME prices in the control solution and then induced fluctuations around the trend. Use of different magnitudes of deviation from trend also failed to reveal any significant nonlinearities.

The lack of any response to fluctuations is the result of a relatively high degree of insulation of the rest of the economy from the export sector, because of either the enclave nature of the sector itself or the effectiveness of countercyclical policy, or both. We did not incorporate any deliberate policy actions into the assumptions, so the results have to be caused by the lack of a strong direct link between the copper-export sector and the rest of the economy. The impact of fluctuations in export revenues was borne by profits, and thus partly by foreign shareholders, providing the Zambian economy with a further cushion.

We have not made any assumptions about deliberate countercyclical fiscal or monetary policies. That is the subject of the next two chapters. The lack of any substantive impact on macro variables could therefore be the result of a relatively isolated export sector or macro behavioral patterns that respond only to long-run trends, or a combination of both. The net international reserves certainly fluctuate with copper prices, as do government revenues. The lagged response of government-investment expenditures, however, is such that these effects are smoothed out, particularly since no asymmetries are incorporated in these functional relationships. Part of the reason also lies in the mining industry's ownership arrangements, whereby the foreign partners' share of the dividends is a residual that absorbs some of the shock from fluctuating profits. The overriding influence, however, seems to be the lack of a substantive direct link between copper mining and the rest of the economy, particularly agriculture and manufacturing.

Note

1. See T. G. Priovolos, "An Econometric Model of the Ivory Coast," a report prepared for a project of Wharton Econometric Forecasting Associates, Inc., and the Agency for International Development, 1979. Models for the other nations were constructed by different individuals at Wharton EFA, Inc., working almost simultaneously.

Appendix 5A: Variable Definitions for Solutions Assuming Random Price Fluctuations

CUQCZM	Copper Production (in thousands of metric tons)
IGFC65	Private Fixed Investment at 1965 prices
LEIND	Total Employment in all industries of Zambians and non-Zambians (in thousands of man-years)
PLCME	London Metal Exchange Price of copper (in kwacha per metric ton)
NINTR	Net International Reserves (in millions of kwacha)
PC	Consumer Price Index
PCDP	GDP Deflator
WGAMQZ	Average Annual Earnings of Zambians in mining and quarrying (in kwacha)
XAFFS65	Output from agriculture, forestry and fishing (in millions of kwacha) at 1965 prices
XGDMC65	GDP at 1965 prices

Policy Issues in Zambia

This chapter gives a brief discussion of the major economic goals of the Zambian society and of how government policy has been used to attain these goals. In chapter 7, some of the policies and their effects are analyzed using the model presented in previous chapters. In this chapter, we discuss basic problems that arise either because of the dual nature of the Zambian economy and its development needs or because of the impact of the copper-export sector on the rest of the economy.

The basic macroeconomic goals of developing countries and the impact of commodity exports on these goods have been discussed elsewhere.[1] These include a rapid increase in per capita income or product, full utilization of productive capacity, nominal stability, and the domestic distribution of income.[2] The structure and circumstances of the economy alter somewhat the nature of these goals when applied to Zambia. For example, capacity utilization is a problem only when unemployment or underemployment of labor is considered; there does not seem to be underutilization of capacity in manufacturing, except when it is caused by raw-material shortages or transportation bottlenecks. Also, the problems of income distribution are compounded by the presence of a highly skilled, highly paid expatriate labor force, which adds a racial and foreign-exchange dimension to the urban-rural income differential.

The major goals and problems that have at one point or another been articulated by the Zambian government are (1) the desirability of balanced economic growth, both by region and by industry; (2) a more diversified basket of commodity exports, so that the country is less dependent on copper for foreign-exchange earnings; (3) a more rapid rate of mineral development, particularly the development of new mineral deposits, and factor proportions that are relatively biased in favor of local labor; (4) provision of capital and credit to Zambians; and (5) a more influential position in Southern African affairs, which is associated with being less dependent on South Africa in economic matters. We consider each of these goals in turn and what part the copper industry plays in its attainment, as well as the relevant policy decisions taken by the Zambian government.

Balanced Economic Growth

The government of Zambia has been concerned with the regional and sectoral imbalance in the country's development.[3] The country is divided into eight provinces. Table 6–1 presents some measures of the regional distribution of economic activity. The Copperbelt Province, for obvious reasons, has the largest population, the highest per capita income, and the largest number of manufacturing concerns. The Luapula, Northern, Western, and Northwestern Provinces are largely rural, relatively poorer, and for years have supplied the mines with migrant labor. The Central Province contains the capital city as well as the location of the central government, and it, too, has attracted a substantial proportion of industry and migrant labor.

The benefits from the mineral industry do not seem to have been evenly shared by all the provinces. If the regional inequities cause problems, then the alternative seems to lie in agricultural development of the rural provinces. The available information shows that nearly all the commercial farmers are located in the Central and Southern Provinces.[4] Agricultural development of the poorer provinces would involve increasing yields through commercializing the subsistence farmer in terms of production methods. In our model, that would involve introducing capital as an input.

The regional imbalance has been accompanied by a substantial variation in sectoral growth rates since independence (see table 6–2), with rates from −2 percent (mining) to 7.7 percent (manufacturing). The lack of growth in the subsistence sector has to have serious consequences, since 62 percent of the population is dependent on this sector's output. The result may have been an aggravation of the urban-rural income differentials.

In 1972, the government introduced the concept of intensive development zones (IDZ) into its development planning. This involves the selection of growth points in the rural areas, based on potential crop returns from incremental use of biological inputs, rural population density, and availability of infrastructure. The zones and projects are province-specific and involve an inflow of foreign exchange. There is not enough information to allow for the evaluation of this attempt at decentralization, all of which was designed to encourage rural development.

Table 6-1
Output per Capita, by Province, 1964 and 1970
(kwacha)

Province	*1964*	*1970*
Northern	12.0	29.0
Copperbelt	264.0	322.0
Eastern	14.0	30.0
Central	71.0	132.0
Southern	37.0	56.0
Luapule	12.0	24.0
Northwestern	13.0	35.0

Source: Republic of Zambia, *First National Development Plan, 1966-1970.* Office of National Development and Planning, Lusaka, July 1966.

Table 6-2
Growth of Real Domestic Product, 1965-1976
(percentage per annum)

	1965-1970	*1970-1976*	*1965-1976*
GDP	2.00	2.80	2.40
Agriculture and forestry	1.70	3.60	2.70
Subsistence	0.70	1.00	0.90
Commercial	5.90	9.10	7.70
Mining and quarrying	-4.80	0.30	-2.00
Manufacturing	11.20	4.80	7.70
Construction	-2.40	4.50	1.40
Transportation and communications	2.70	1.10	1.80
Services	9.30	4.50	6.70

Source: Republic of Zambia, *Monthly Digest of Statistics,* Lusaka, several issues.

The problems of an uneven growth pattern are not likely to be resolved soon, simply because of Zambia's reliance on mineral production and the concentration of those minerals in a relatively small geographic area in the northwest. While plans call for an expansion of mineral production, agriculture seems to be stagnant. At the time this book was being written, there were plans to expand cobalt production from Chibuluma mine by 1980. Declines in output from agriculture and slow growth in manufacturing continued to plague the country into 1980, resulting in further imports. Development of steel production would significantly aid the capital-goods industry. There are sufficiently large iron-ore deposits near Lusaka, but a project to develop them has been beset by problems.

Composition of Exports

Both the First and Second National Development Plans have noted the reliance of Zambia on a single export commodity and the need to diversify the base. Exports of manufactured goods are unlikely to develop in the foreseeable future, and the only alternatives are exports of other minerals or agricultural commodities. Both require foreign exchange, which is in short supply and can only be earned by copper exports. The problem of how to reduce the dependency on copper has thus been a difficult one, since it requires copper-export earnings for investment into the production of alternative exports or import substitutes. Most of the projects undertaken by the state's Industrial Development Corporation (INDECO) since its formation have been designed to replace traditional imports, including textiles, glass and industrial fibers, and basic metals. Alternative exports (aside from lead and zinc) have included cobalt and semiprecious stones. These face the same production and transportation problems as copper. Expansion of agricultural exports and import substitutes is beset by the same problems as with other crops.

State ventures into large-scale beef and sugar production have been hampered by the shortage of imported inputs and managerial skills.

Mineral Development

As has been discussed, the government of Zambia would like to see the mineral industry expand, preferably through the opening of new mines. Zambia would also like to maximize the returned value of mineral output, consisting of government tax receipts, dividends, and payments to intermediate inputs. This has been done by trying to increase employment of domestic labor and by designing a tax system that captures as large a share of net profits as possible during a price upswing while insulating the tax receipts against downswings. The goals have not been compatible. Expansion of output will probably come at the expense of the relative share of domestic inputs in total costs, because the incremental output would be from relatively capital-intensive operations. Since independence, the composition of output has changed in favor of open-pit operations. Production from the major underground mines, which employ relatively more labor per unit of output, has declined, while open-pit operations and the processing of tailings, both relatively capital-intensive, have increased production. The trend could be the result of technical considerations, relative-factor prices, or the probable existence of economies of scale in open-pit production.

It is beyond our scope to determine the importance of technical considerations in the choice of production methods and whether there is substantial flexibility. Once the method of production was selected, however, there seemed to be a great deal of substitutability of inputs, particularly in the actual mining of ore. Until 1972, six different methods were used to extract ore in underground mining, and each implied different factor proportions.[5] The trend was for most of the production to come from capital-intensive open-stopping and block-caving methods. By 1972, for example, the two methods accounted for 70 percent of the ore extracted at Mufulira mine. The reason given for this trend was the consistent increase in mining-labor costs.[6] The capital-depreciation laws introduced in 1968 biased factor proportions further in favor of capital. They allowed for any investment in earth-moving equipment to be totally written off in the year it was made for tax purposes. The intention was to encourage the development of new mines to increase both output and employment. The result, however, was to raise the marginal capital-labor ratio at existing mines. The spread of trackless mining accelerated at all mines, and the purpose was to improve profitability.[7] The laws were repealed in 1975, but the effects on factor proportions seem to have remained.

This conflict between increasing outputs, profits, and employment has troubled the government of Zambia since independence. Increases of both government revenue and output depend on minimizing costs. Social and union pressures to increase Zambian wages have resulted in labor costs that are higher than can be justified by unskilled-labor-supply conditions alone. Declining ore grades have forced the industry to mine more ore per ton of finished production, and the locations of remaining deposits and tailings seem to favor open-pit mining. It is more efficient in mining poorer ores, but it also generates less incremental employment.

Our model is not detailed enough to deal with the impact on GDP of changing mining processes. However, one could estimate the impact of declining ore grades on labor utilization, given the mining method used. The same can be done to assess the impact of price changes on output and employment, given certain assumptions about government policy on wages or the mining method used to obtain the incremental output.

A related concern is that of maximizing the employment of domestic labor by replacing expatriates with skilled local employees. The supply of skilled local labor has been the constraint to this approach. Some Zambianization has occurred; by 1978 the toal number of expatriate employees was about 3,600, down from 7,000 in 1965. These are one-shot gains, however. Once the pool of replaceable expatriates is exhausted, any further gains would have to come from increases in production. The training of local labor, which is exogenous to our model, would still be the binding constraint.

One could argue that, historically, Zambian mining policy has not necessarily focused on profit maximization. During periods of high prices (1964–1966, 1973–1974), the government has concentrated on increasing the country's share by cutting into foreign shareholders' dividends. When prices were low (1970–1971), the emphasis was on reducing the foreign interest itself and increasing the domestic share of costs. This would suggest a policy aimed at maximizing the domestic share of total mineral-export value, with either profits or employment, or a suitably weighted combination of both, as the objective function. The idea of an objective function with both profits and employment in it may be a reasonable alternative. If wages for local labor were constant, then maximizing employment would be the same as maximizing the wage bill for domestic labor. The objective function would then be maximization of the domestic share of the value of output, consisting of tax revenues and local wage payment, each suitably weighted to account for the fact that a unit increase in domestic labor payments implies a decline in tax revenues. Our model assumes profit maximization, but we can make assumptions about employment targets, given prices, and then simulate the impact on the rest of the economy.

Provision of Capital and Credit to Zambians

Until 1964, the country's financial institutions rarely granted credit to the African population, particularly the Zambian small farmer. The reasons were never articulated but probably had to do with the standards used to assess credit-worthiness. The ordinary African farmer did not own his land and therefore could not use it as collateral. African businessmen could not own property in the so-called first-class trading areas, and this probably increased the degree of risk financial institutions assigned to their investments.

Whatever the reasons, the government of Zambia believed the financial institutions had been discriminating against Zambians,[8] particularly those who had no credit history. The initial response was to limit the ability of expatriates to borrow in the domestic market by subjecting their loan applications to foreign-exchange controls. Any loans to expatriates had to be accompanied by an equivalent inflow of foreign currency. At the same time, the government delineated certain geographic areas and businesses as being open to local entrepreneurs only. The assumption seems to have been that expatriates were crowding out local borrowers from the credit market and thus had a competitive edge in business.

State participation in enterprises owned by expatriates seems to have negated the regulations, because the enterprises were then able to borrow on the domestic money market without any restrictions. Once nationalized, the copper-mining industry borrowed rather significant amounts from the Bank of Zambia for the first time; borrowings increased from 2.8 million kwacha in 1970 to 80.2 million kwacha by 1977. Commercial bank loans to agriculture increased from 4.4 million kwacha to 10.4 million kwacha during the same period. Declines in copper-export earnings also caused a government deficit during 1970–1973 and 1975–1976, forcing the government to enter the domestic money market to finance its expenditures. During those periods, economic activity in the private nonmining sector was also sluggish, and the banks still had excess liquidity. Credit availability was not a problem during 1973, when copper prices were high. It is thus not clear that credit availability has been a major constraint since 1971, although financial institutions could conceivably still be very cautious in making loans to black Zambian entrepreneurs.

Position in Southern Africa

Zambia's position in Southern Africa is a difficult issue to handle within the context of a formal model, yet it is an extremely important consideration for the Zambian economy. The internal, racial, and political policies of South Africa are well known and have resulted in that country's exclusion from most international organizations, among them the Organization of African Unity (OAU). Prior to 1979, the same conditions applied to Southern Rhodesia, Zambia's neighbor to the south. Zambian government policy has been dictated partly by domestic needs and partly by the commitment to the edicts of the OAU and the United Nations. The country's effectiveness as a member of the five "frontline states"[9] thus depended on its ability to minimize its economic dependence on the embargoed Rhodesian economy and the more powerful South Africa while increasing the ties with African countries to the north. This policy presented problems in transportation routes for foreign trade, emergency food supplies, imported inputs, and military strength.

Between 1965 and 1978, the major role of these states was to assist the forces fighting the Rhodesian regime by providing them with a haven and a conduit for arms. They also had to abide with United Nations economic sanctions against Rhodesia. Zambia was most affected by this responsibility. As was discussed in chapter 1, the country had to find alternative routes for its foreign trade and even had to resort to using South African ports. Rhodesian industry had provided a significant share of consumer and industrial goods, and these now had to be imported from elsewhere. When the agricultural crop failed in the 1971–1972 season, however, Zambia had to import from Rhodesia at high cost in terms of both prices and prestige. The Rhodesian and South African governments, on their part, tried to prolong this dependence by occasionally destroying parts of the Zambian transport infrastructure. In 1979, the capacity of the Tanzam railway was reduced when Rhodesian forces destroyed a key bridge, and by mid-1980 the damage had not been fully repaired. It is conceivable that the Tanzam railway itself may become a high-cost route once the railway through Zimbabwe is operational. However, Zambia will continue to need alternative routes to the sea, especially if the internal unrest in Namibia (Southwest Africa) intensifies and the much stronger South Africa retaliates on a much larger scale. The effect would certainly be to raise mining costs and general price levels. At a recent meeting, the frontline states designated the new nation of Zimbabwe as having the responsibility of insuring the region's food supplies. Since the weather patterns in Zambia and Zimbabwe are similar, it is unlikely that the resolution will significantly change Zambia's food picture; in any case, Zambia would still require foreign exchange for the purchases. There were also discussions about a regional program to improve the transportation system, including the port of Beira in Mozambique, institutions to train mining and engineering personnel, and an information network, all independent of South African sources.

These political considerations loom very large in the Zambian economy. Our model could be used to test for their impact, since they are likely to affect copper production, exports, transportation costs, foreign-exchange reserves, or the domestic price level. The events are likely to be sporadic, and their impact could be tested by inducing a one-period change in the relevant variable while holding all else constant.

Price Stability

Government concern with price stability manifests itself in the system of producer prices for agricultural commodities and industry wage-price guidelines.

Prices paid to farmers for controlled commodities are set by the various marketing boards. Although producer prices are supposed to vary depending on supply conditions, it would seem that stability is an overriding concern. Thus, producer prices for maize remained almost constant between 1964 and 1974 and actually fell in real terms, although there was a shortage during 1970–1972. Similarly, producer beef prices were kept constant, with the government paying a subsidy to millers of stock feed. Maximum retail prices for major food commodities are established by statute. Changes in the consumer price index, therefore, have largely been caused by imports. The impact of these controls, of course, may have been to curb domestic production, thus stimulating imports and further price increases. Shortages of beef and other food commodities have plagued the country intermittently since 1970, and price controls may have played a significant part.

The controls apparently were designed to keep urban food prices reasonable and thus curb wage increases. When wages for mining labor continued rising, however, the government instituted three-year, fixed-period wage agreements; the first two covered 1970–1976. Increases were limited to 5 percent per year.[10] The agreements virtually eliminated industry-wide strikes during the period covered but may have institutionalized wage increases that were not necessarily justified by market considerations. In any case, the combination of fixed agricultural producer prices, stagnant agricultural production, and steadily rising mining wages has prevented any reduction in the urban-rural income gap, particularly since wage employment for Zambians was growing at just over 2.5 percent during 1966–1976.

Our model shows that a change in wages does not directly affect employment in most of the industries. In mining, wages would affect the real price of copper as well as profits. These, in turn, affect total output, investment, and thus employment, not only in mining but also in other sectors. Another policy alternative would be to let agricultural producer prices rise to the level of import prices while keeping wage incomes constant, and for this increase to be reflected in the consumer prices. Real wages in all industries would decline, but mining costs would not be affected. Agricultural production should respond, given the price elasticities of 0.28 in the commercial sector and 0.033 in the subsistence sector.

Provision of Skilled Manpower

Zambia has been plagued by a shortage of skilled manpower. The government would rather have its own citizens trained for skilled positions, since expatriate labor involves foreign exchange and search costs that are not applicable to labor of domestic origin. There is also a feeling that expatriates are not fully committed to the country, which affects their productivity.[11] As discussed in chapter 1, however, higher education for Africans was practically nonexistent before 1964, and no Africans were in the skilled-trade and management positions. There was therefore a conflict between the desire to maintain productivity of the industry in the short run, using expatriate labor, and training Africans to fill the skilled positions. The mining industry had its own training schemes and also provided scholarships for studies overseas and at the new University of Zambia. The transition process has not been smooth. The net rate of exodus of the expatriate labor force has been higher than the increase in the number of skilled Zambians to replace them. The government at first paid the management of the industry a recruitment fee to provide the skilled manpower, even as a selective employment tax was imposed on salaries paid to non-Zambians. In 1977, the government imposed an annual education levy of 120 kwacha on every corporation. The proceeds are to be used to finance education and training of skilled labor.

Since 1972, the country has not been able to reduce the absolute number of expatriates; the numbers in mining have actually increased slightly. The problem seems to be the shortage of Zambian high school leavers with the necessary mathematical or science background.[12] The model does not explain the availability of local skilled labor. It is possible that Zambians would command a lower wage than would expatriate labor with similar skills, especially since the Zambians do not have to be paid an expatriate bonus. The supply of skilled local labor, however, is considered exogenous to the model. Nevertheless, we can take this into account by increasing government expenditures on social services (including education) and then making some assumptions about the output of local skilled labor. The amount of expatriate labor could be reduced accordingly, with or without an impact on the total wage bill.

Notes

1. See J. R. Behrman, "General Considerations on Model Specifications for Integrated Econometric Analysis of International Primary Commodity Markets and Goal Attainment in Developing Countries," unpublished manuscript, Wharton Econometric Forecasting Associates, Inc., Philadelphia, 1978.

2. Ibid.

3. See Republic of Zambia, "Zambia's Economic Revolution," an address by the president, Dr. K.D. Kaunda, at Mulungushi, 19 April 1968, pp. 14–15.

4. Republic of Zambia, *Second National Development Plan*. Ministry of Development Planning and National Guidance, December 1971, pp. 298–301.

5. The methods were open stopping, mechanized open stopping, mechanized strike retreat, cascade stopping, block caving, and sublevel caving, in increasing order of capital-labor ratios.

6. Roan Selection Trust.

7. Trackless mining involves the use of self-loading trucks to mine ore and to transport it to the surface, as opposed to the combination of shovels and rail trucks.

8. The definition of "Zambian" in this case means someone holding a Zambian passport or green national registration card, or a corporation wholly owned by Zambians. See "Zambia's Economic Revolution," pp. 28–30.

9. The frontline states are Mozambique, Zambia, Tanzania, Botswana, and Angola. Now that it is independent, Zimbabwe is likely to join the group as the sixth member.

10. See *Minedco Mining Yearbook of Zambia,* 1970, 1973.

11. See "Towards Complete Independence," an address by the president to the United National Independence Party National Council, August 1969, p. 9.

12. See *Zambia Mining Yearbook 1977,* Copper Industry Service Bureau, Zambia, p. 18.

Policy Analysis and Simulations

Introduction

The previous chapter discussed the goals of the Zambian society and associated policy issues. In this chapter, we deal with specific domestic economic policies that could be applied to offset the impact of copper-price fluctuations and policies that promote the attainment of the broad macroeconomic goals discussed in the last chapter. The macroeconomic goals we consider include the growth of real income, employment of Zambians, and the stability of domestic prices and output. The policies considered for use in offsetting the impact of copper-price fluctuations are the traditional policies of macroeconomic management, namely, fiscal and monetary policies.

The other policies considered are devoted to specific sectors of the Zambian economy, such as commercial policy in the copper industry and policies for agricultural development. Copper-industry activities affect the stability of export and government revenues, the balance of payments, gross domestic product, employment, and price stability, and so they also have implications for the broad goals. Developments and policies in agriculture have impact on income distribution, migration, domestic prices, regional growth and income disparities, diversification of exports, and import substitution, and thus have implications for balanced growth. The impact of the general economic and sector-specific policies will be explored by simulations with the model of the Zambian economy.

In the analysis that follows, we distinguish between a passive-policy scenario, which is usually assumed in the control solution, and an active-policy scenario. The passive-policy scenario assumes that the policymaker, usually the government, behaves in the fashion embodied in the parameters and structural characteristics of the model of the economy. An active-policy scenario is one in which the policymaker changes its decision rules in response to a new policy environment or moves in a new policy direction. For example, when the price of copper rises, the parameters of the model indicate that the Central Bank sterilizes some but not all the changes in foreign-exchange inflows. A policy that sterilizes all or none of the inflows is an active policy, in contrast to a passive policy, which follows the usual rule and sterilizes only some of them. In the simulations that follow, we compare solutions produced by assuming a particular active policy with a solution embodying passive policy.

Monetary Policy

In the Zambian economy, monetary policy is concerned with the supply of money, the control of credit, and the setting of rates of return on financial instruments. In a developed economy, money supply is presumed to affect economic activity through changes in interest rates, which affect investment and consumption. Money supply may also affect the price level, depending on the degree of slack in the economy. For an economy such as Zambia's, with a small and highly regulated capital market, a thin pool of skilled and experienced labor force, no capital-goods industry, and a low per capita income, the effects of monetary policy will be different from those of a developed economy. The fact that almost all capital goods will have to be imported, and the lack of a large pool of trained and skilled labor force, places a severe constraint on short-term changes in aggregate output. The existence of a nonregulated informal capital market reduces and distorts the effectiveness of monetary policy. Thus, monetary policy cannot be an effective tool to fine tune the economy. In the short run, increases in money supply lead to inflationary pressures generated by excess demand. Since demand responds faster than supply in Zambia, monetary policy may therefore be a potent tool for the control of inflation rather than of output. The supply of credit also influences gross investment, so money supply has consequences for economic growth. There is therefore a trade-off involved in the determination of the optimal supply of money.

In the Zambian economy, the money supply is influenced by the level of copper prices through its effect on foreign-exchange earnings and government revenues. The fluctuation of copper prices results in fluctuations in the money supply unless the Central Bank is able to sterilize foreign-exchange flows. Fluctuation of money supply leads to unanticipated demand-induced inflation, which makes the control of money supply crucial for price stability. The goal of monetary policy should then be to provide the economy with sufficient liquidity to finance current transactions and to provide capital for investment, but it should not result in undesirable levels of inflation. Monetary policy should then attempt to insulate the financial sector of the economy from the effects of the short-term fluctuations of copper prices. It should be concerned with long-term growth and price stability.

The responsibility of carrying out the Zambian government's monetary policies lies with the Central Bank of Zambia. The Central Bank is responsible for controlling the money supply, for managing the public domestic and external debt and assets, for foreign-exchange control, and for overseeing the operations of commercial banks. It is also responsible for raising foreign loans on behalf of the government and parastatal firms. In addition, the Central Bank lends money directly to parastatal corporations and to the mining industry, usually in conjunction with other financial institutions. It also holds deposits of the mining companies as a way of controlling the monetary impact of foreign-exchange flows.

The major instruments of controlling the Zambian money supply are (1) direct control of the credit operations of commercial banks; (2) control of foreign-exchange flows; and (3) manipulation of the required reserve ratios and rediscount privileges of commercial banks. Traditional open-market operations, that is, buying and selling by the monetary authority of government financial instruments, have limited effectiveness for money-supply control in the Zambian economy. The commercial banks are the principal buyers of the treasury bills and government bonds, and these instruments can be used to meet legal reserve requirements. In addition, Zambian banks usually maintain liquid reserves far in excess of the legal requirements.

It was shown in chapter 5 that the greatest source of changes in the money supply is the effect of fluctuations of copper prices on foreign-exchange earnings. Another source of endogenous money-supply changes is the change in government borrowing from the Central Bank to finance deficits. This, too, is partly. induced by changes in government revenues engendered by copper-price volatility. One way of insulating the money supply from these fluctuations is to set targets for the growth rate of money supply, based on long-term considerations. The Central Bank could then use its monetary tools—credit control, reserve requirements, sterilization of foreign-exchange flows, and so forth—to ensure that the money supply grows according to target.

Simulation Tests

The effects of monetary policy on the Zambian economy will be investigated with the model, on the assumption that the Central Bank can effectively control the supply of money. In chapter 5 we reported historical simulations in which exogenous copper-price and output shocks were applied to the economy represented by our model. In one simulation, the model was solved with a sustained 10 percent increase in historical copper prices, and we found that this resulted in higher levels of aggregate domestic prices and increased real domestic output.[1] Here we solve the model with a sustained 10 percent increase in copper prices, but in this case, the Central Bank sterilizes the induced inflows of foreign exchange. The money supply is exogenized and the values are held at those of the base solution. The results of the two simulations—passive monetary policy, embodying endogenous money supply, and active monetary policy, which holds the money supply fixed at base-solution values by sterilizing the impact of foreign-exchange flows—are compared in table 7–1.

The results indicate that the money supply in the passive-policy solution is much higher than with the active policy (as expected), and, as a result, the price level is higher for the passive-policy solution. The difference in total output is very small, being higher for the solution where the money supply is exogenous, with most of the increase coming from the services sector. The lower money supply in the exogenous money solution results in the fall of noncopper private investment, although investment in copper rises for this simulation. The rise in copper investment is a result of increased profits arising from lower input costs. Public

Table 7-1
Effect of a Sustained 10 Percent Increase in Copper Prices with Money Supply Held Exogenously at Base-Solution Values

	Percentage Differences between Solution with Exogenous Money Supply and Solution with Endogenous Money Supply								
Variables	*1968*	*1969*	*1970*	*1971*	*1972*	*1973*	*1974*	*1975*	*1976*
Real GDP	0.18	0.25	0.29	0.38	0.33	0.21	0.20	0.14	0.07
Agriculture	0.00	0.00	–0.02	–0.03	–0.05	–0.07	–0.09	–0.11	–0.11
Mining	0.01	0.02	0.04	0.04	0.04	0.05	0.05	0.03	0.05
Manufacturing	0.00	–0.02	–0.09	–0.20	–0.28	–0.44	–0.58	–0.70	–0.75
Services	0.31	0.51	0.60	0.72	0.68	0.58	0.57	0.50	0.40
Private consumption, real	0.24	0.51	0.54	0.49	0.41	0.34	0.30	0.07	–0.20
Public consumption, real	0.77	1.27	1.41	1.61	1.61	1.41	1.41	1.16	1.00
Real gross investment	0.11	–0.39	–0.58	–0.47	–0.72	–1.24	–1.12	–0.91	–1.28
Investment in copper	1.00	2.51	2.68	2.19	2.14	2.00	1.96	1.86	1.78
Other private investment	–1.05	–1.64	–2.05	–2.73	–2.79	–3.65	–3.57	–2.17	–2.58
Total urban employment	0.12	0.22	0.29	0.34	0.33	0.27	0.27	0.23	0.17
Total wage bill	–0.42	–0.88	–1.04	–1.26	–1.49	–1.42	–1.28	–1.11	1.17
Operating surplus	–0.25	–0.35	–0.39	–0.80	–0.75	–0.71	–0.90	–1.30	–1.34
Money supply	–5.45	–8.71	–9.92	–11.10	–10.90	–9.72	–10.20	–8.45	–7.53
GDP price deflator	–0.49	–0.80	–1.09	–1.44	–1.49	–1.23	–1.24	–1.36	–1.28
Consumer price index	–0.75	–1.29	–1.51	–1.74	–1.77	–1.59	–1.58	–1.33	–1.20
Government current expenditure	–0.06	–0.24	–0.41	–0.52	–0.68	–0.72	–0.66	–0.59	–0.62
Public direct investment	–0.06	–0.36	–0.58	–0.80	–1.22	–1.38	–1.14	–1.11	–1.57
Government current revenue	–0.31	–0.63	–0.81	–1.39	–1.86	–1.51	–1.10	–1.82	–2.00
Imports of goods and services	0.14	0.02	–0.09	–0.03	–0.17	–0.17	–0.33	–0.34	–0.58
Exports of goods and services	0.01	0.02	0.04	0.04	0.05	0.05	0.05	0.04	0.04

real consumption is higher for the active-policy simulation, since nearly the same amount of nominal expenditures is made but at lower costs. The results would support the conclusion that monetary policy has a negligible impact on real economic activity but a significant effect on the aggregate price levels.

The second monetary-policy test is to examine the effects of a constant growth rate of money supply. This type of policy is usually associated with the protagonists of the quantity theory of money and is actively advocated by them for developed economies and even more so for developing economies.[2] The major argument for a stable or constant growth of money is that, in the long run, the money supply determines only the inflation rate, and a steady and anticipated inflation rate is preferable to an unsteady and unanticipated inflation. In the Zambian economy, money-supply changes do not appear to have much significant impact on economic activity in the short run; in the long run there appears to be a trade-off between the inflationary impact and its impact on growth through investment. A stable growth rate of money supply may also offset the effect of copper-price fluctuations and reduce their destabilizing impact.

To test the effect of a stable growth in the nominal money supply, we exogenize the money supply in the model and apply various growth rates close to the compound monetary growth rate of the control solution, also keeping the average money supplies in all the solutions close. The copper prices used in these simulations are generated randomly, with a mean along a stable growth path and the same standard deviation as the historical copper prices. The solution is performed within and outside the sample period—from 1968 to 1990; the solution in which the money supply is endogenous is the control solution. The terminal value, means, and trend-adjusted standard deviations of selected variables from the control solution, and solutions with exogenous constant growth rates of money, are shown in table 7–2. The control solution represents the economy with passive monetary policy, and the other solutions, with exogenous money, represent active monetary policy. The standard deviations are the deviations of the logs of the variables from log-trend values.

The results indicate that the higher the growth rate of money, the higher the gross domestic output, as measured by both the means and the terminal values. With the exception of the mining and quarrying sector, the value added of the other sectors increases with higher growth rate of money, with the highest increase being from the construction sector (not shown in the table). The aggregate price level, as measured by the consumer price index, also increases with higher growth of money supply.

The terminal values of the money supply in the constant-growth solutions are higher than the value in the control solution, but the average values are lower. The money supply in the control solution moves up and down, following changes in copper prices. With the exception of the lowest-growth money supply, the terminal values of the gross domestic output are higher for the constant-growth money solutions than for the control, although the average values of real GDP are slightly

Table 7-2
Impact of Different Growth Rates of Money Supply under a Regime of Fluctuating Copper Prices, 1968-1990

	Endogenous Money (Growth Rate = 11.8% per annum)			*Exogenous Money (Growth Rate = 11.95% per annum)*		
Variables	*Terminal Value*	*Mean Value*	*Standard Deviation*	*Terminal Value*	*Mean Value*	*Standard Deviation*
Real GDP	1452.5	1071.4	0.0483	1448.3	1066.6	0.0507
Agriculture	190.2	148.8	0.0224	189.0	148.1	0.0221
Mining and quarrying	176.9	184.9	0.1670	176.6	185.0	0.1667
Manufacturing	184.4	135.1	0.0804	181.5	133.2	0.0787
Services	726.0	467.6	0.0382	725.0	468.4	0.0400
Private consumption, real	535.0	435.5	0.1071	521.8	422.4	0.1166
Public consumption, real	345.0	245.5	0.0919	343.7	246.7	0.9603
Real gross investment	402.7	309.2	0.1437	406.1	294.8	0.1625
Total urban employment	508.8	416.3	0.0255	507.2	416.0	0.2702
Money supply	2955.6	1367.2	0.1040	3051.8	1150.4	0.00
GDP price deflator	4.344	2.835	0.0867	4.357	2.812	0.0877
Consumer price index	199.4	138.0	0.1104	200.6	134.8	0.1061

	Exogenous Money (Growth Rate = 12.5% per annum)			*Exogenous Money (Growth Rate = 13.0% per annum)*		
Variables	*Terminal Value*	*Mean Value*	*Standard Deviation*	*Terminal Value*	*Mean Value*	*Standard Deviation*
Real GDP	1455.7	1068.4	0.0510	1463.0	1070.1	0.0513
Agriculture	189.6	148.2	0.0220	190.1	148.3	0.0220
Mining and quarrying	176.8	185.0	0.1667	176.8	185.0	0.1668
Manufacturing	182.6	133.5	0.0783	183.7	133.8	0.0779
Services	725.7	468.3	0.0398	726.2	468.3	0.0397
Private consumption, real	532.8	425.7	0.1178	543.5	428.7	0.1191
Public consumption, real	344.5	246.6	0.0958	345.4	246.6	0.0956
Real gross investment	423.7	299.9	0.1643	441.4	304.8	0.1662
Total urban employment	508.1	416.1	0.2698	509.2	416.2	0.0270
Money supply	3399.3	1241.6	0.0	3747.5	1331.1	0.00
GDP price deflator	4.358	2.816	0.0878	4.358	2.820	0.0879
Consumer price index	202.4	135.7	0.1065	203.9	136.5	0.1069

less for the constant-growth solutions. Similarly, the terminal values of the domestic price indexes are higher for the constant growth solutions than for the solution, but in all cases the average price levels are lower than for the control.

The deflator for gross domestic product (PGDP) appears to have the same terminal values in all three constant-growth solutions. Part of the reason for this is that rounding to the third decimal place obscures any differences; the values are different but the differences are small. The higher the money-supply growth rate, the higher the GDP deflator. A more fundamental reason for the closeness is that deflators of two major components of gross domestic product—mining and agriculture—are invariant to changes in the money supply. Mining deflator responds to copper prices and agricultural deflator responds mainly to exogenously set agricultural producer prices. Another contributing factor is that increases in money supply increase noncopper investment, so that, in the long run, noncopper output is increased. This induced rise in output tends to counter excess-demand-induced inflation.

The trend-adjusted variance of gross domestic output, final demand components, and employment are higher for the stable-growth-rate solutions. The variance of the price level as measured by the consumer price index is lower, but that of the deflator for GDP is slightly higher for the stable-growth-rate solutions. This simulation test was repeated using a copper-price series with a higher variance, which, when the money supply is endogenous, produces a variance of money supply that is higher than in the control solution. Comparing an endogenous money solution with a stable-growth-rate money solution produced the same pattern of variance as in the first test. The variance of the price level was lower in the stable case, while most other major variables had higher variances. The results of these tests would indicate that stabilizing the nominal money supply alone stabilizes only the price level but does not reduce the instability caused by copper-price fluctuations in output, investment, consumption, and employment. Later in this chapter we shall provide an explanation of why stabilizing the money supply does not lead to real stability.

Fiscal Policies

Zambian government-spending decisions are constrained by the financial and physical resources available to the government. The financial resources are influenced by the price of copper; the fluctuations in copper prices result in the volatility of government revenues. Changes in the price of copper induce the government to alter their spending plans. For example, the fall in copper prices in 1975 led to several reductions in the 1976 and 1977 budgets. In 1976, subsidies and net capital expenditures were drastically reduced from their 1975 levels. Jolly and Williams, also report that cuts in expenditures were introduced to the 1972 budget following the decline in copper prices in 1970 and 1971.[3]

One of the objectives of the Zambian government is to free its current and capital spending from being dictated by short-run price fluctuations. One possible way to achieve this is to base spending plans on long-term expectations of financial resources, including copper revenues. Short-term changes in revenues could then be smoothed out by a combination of domestic and foreign borrowing and lending. Spending decisions are one facet of fiscal policy; the other side is the raising of revenues. This is interesting in itself but not nearly as important, in the context of the present work, as spending, which will be the major focus in this analysis.

A government-spending policy that is fixed in the short run is not free of political and economic risks. Politically, it may be difficult for the Zambian government to restrict spending along its long-term target path when revenues are increasing. We tested for any asymmetric behavior in Zambian government-spending decisions and concluded that the data did not support such behavior. However, the period for which this test was performed was one during which physical constraints, such as manpower, were more binding than financial constraints. The economic risks include the loss of credit-worthiness in international financial markets and inflationary and deflationary effects, which may result in borrowing and lending from abroad and home, respectively. Since the copper-price cycle can be long, the government may run up a high external debt, which may impair its ability to borrow or even service and repay any existing loans.

Despite the risks, a policy of insulating government expenditure from the short-term fluctuations of copper prices by fixed spending plans has theoretical appeal.[4] To accomplish this policy it is important (1) that the government have the fiscal discipline to implement such a plan; (2) that effort be devoted to the management of foreign assets and liabilities; and (3) that the plan be based on a realistic assessment of future financial resources. The government should explore all possible sources of loans, including multilateral agencies, discretionary foreign-government loans and grants, private foreign-capital markets, and trade credits, to get the most out of external funds. Management of foreign debt and assets should be a continuing operation that deserves advance planning so that the country obtains the best possible terms for its debts and high returns for its assets. In addition, the riskiness of Zambian debts may be reduced by backing them with a strategic reserve of copper stocks.

Simulation Tests

In the model of Zambia, government spending is in two general categories—capital spending and current spending. The latter is further subdivided into three endogenous categories—general, social, and economic services—and an exogenous residual category. These spending variables are estimated in nominal terms; a common independent variable is lagged government revenues, to which they respond positively (see chapter 4). This implies that spending is reduced, with a lag, when copper prices fall and reduce current revenues. The components of capital spending are direct capital expenditures, which respond to lagged government revenues, and government loans to and investment in the private sector, which respond to the net budget deficit. In the simulations that follow, we manipulate these government-spending variables in an effort to directly or indirectly offset short-term changes in the price of copper.

Offsetting the Effect of a One-Period Decline in Copper Prices. The Zambian government sometimes responded to falling copper prices by cutting its expenditure. This approach tends to reinforce the contractionary effect of a fall in copper prices. Here we test the effect of an active fiscal policy, which attempts to offset the contraction of the economy caused by a decline in copper prices by increasing, not lowering, government expenditures.

The effect of a none-period decline in copper prices by 10 percent was explored and reported in chapter 5. The same test is performed here, but the base-solution values of government expenditures are exogenously increased by 10 percent to counter the effects of the fall in copper prices. To perform the test, four endogenized categories of government expenditure—general, social, and economic services and direct investment—are exogenized at their base-solution values and raised by 10 percent. The solution embodying the exogenous increase in government expenditure and a 10 percent fall in copper prices in 1968 is then compared with a solution in which only the copper price falls in 1968 by 10 percent.[5] The results are presented in table 7–3.

The increase in government expenditure offsets the decline in real output and employment caused by the fall in the price of copper in 1968. With the price of copper reduced by 10 percent from the base-solution value, real gross domestic product declined by 0.6 percent and 0.7 percent in 1968 and 1969 (see chapter 5, table 5–5). When the government expenditure is increased from the base-solution value by 10 percent, the gross domestic product rises by 1.71 percent and 0.56 percent in 1968 and 1969, thus offsetting the effect of the decline in output caused by the fall in the price of copper. Similarly, the decline in employment is offset by the higher government expenditure. The distribution of output in this offset solution is very different from the base solution or the lower-copper-price solution. There is a movement away from the mining sector to other sectors, primarily services, construction, and manufacturing. The offset policy does not have much impact on the domestic price level, which actually declines in the first two years.

The negative aspect of the offset policy is that it increases government domestic and foreign indebtedness, since government has to increase expenditures while its resources are falling. In addition, it depletes the foreign reserves of the country, since an expansionary government budget expands imports at a time when exports are falling. Whether the government is able to carry out this kind of offset policy depends on the amount of both foreign and domestic reserves it has and its capacity to borrow at home or abroad.

Offsetting the Effect of a Sustained Decrease in Copper Prices. In chapter 5, the effect of a sustained decrease in the price of copper was explored. It was found

Table 7-3
Effect of a One-Shot 10 Percent Decrease in Copper Prices and a One-Shot Increase in Government Spending in 1968

	Percentage Differences			
Variables	*1968*	*1969*	*1970*	*1971*
Real GDP	1.71	0.56	0.14	0.45
Agriculture	0.00	0.14	0.15	0.13
Mining and quarrying	0.00	0.00	0.00	0.00
Manufacturing	0.00	1.17	1.16	0.92
Services	4.19	1.26	0.14	–0.15
Private consumption, real	1.05	0.99	0.81	0.65
Public consumption, real	9.84	3.18	0.34	–0.34
Real gross investment	5.26	1.58	–0.43	–0.25
Investment in copper	0.50	0.46	0.11	–0.07
Other private investment	0.21	–1.57	–1.16	–0.05
Total urban employment	1.32	0.58	0.23	0.09
Total wage bill	1.08	0.40	0.18	0.19
Total operating surplus	2.67	0.50	–0.03	–0.04
Money supply	–0.53	0.22	0.44	0.58
GDP price deflator	–0.42	–0.18	–0.01	0.06
Consumer price index	–0.37	–0.12	0.05	0.05
Government current expenditure	9.71	6.84	0.88	–0.54
Government direct investment	11.38	7.76	0.66	–0.72
Government current revenue	2.14	0.53	0.12	0.13
Imports of goods and services	3.32	1.04	–0.16	–0.12
Exports of goods and services	0.00	0.00	0.00	0.00
	Differences			
Government net budget deficit (–)	–31.50	–18.60	0.00	3.24
Balance of trade	–12.00	–3.50	0.50	0.50
Balance of current account	–16.60	–3.20	2.70	2.34
Net international reserves	–12.70	–15.10	–13.10	–11.40

that this event reduced economic activity in Zambia, particularly in the mining industry. We shall test for the effect of a government policy that counters this decline by increasing government expenditures. The test proceeds as in the previous section, except that a sustained 10 percent increase on base-solution values is made to the same categories of expenditures, that is, general, social, and economic services, and direct investment. We note here that this policy involves increasing government expenditures above their normal (base-solution) levels at a time when passive policy would have led to a reduction of expenditures below the base-solution levels.

The results of this active-offset-policy solution are compared with the passive policy in table 7–4. This table should be contrasted with table 5–10 in chapter 5, which compares the base solution with one shocked by a sustained 10 percent decrease in the price of copper. Table 7–4 shows that the increase in government

Table 7-4
Effect of a Sustained 10 Percent Decrease in Copper Prices and an Autonomous Sustained 10 Percent Increase in Government Spending

Variables	1968	1969	1970	1971	1972	1973	1974	1975	1976
	Percentage Differences								
Real GDP	1.71	2.11	2.57	2.45	2.12	2.16	2.32	2.46	2.16
Agriculture	0.00	0.14	0.25	0.30	0.32	0.31	0.32	0.34	0.35
Mining and quarrying	0.00	0.00	−0.01	−0.01	−0.02	−0.02	−0.02	−0.00	−0.02
Manufacturing	0.00	1.17	1.61	2.06	2.09	2.07	2.16	2.34	2.34
Services	4.19	5.32	5.87	5.29	4.51	4.32	4.51	4.74	4.04
Private consumption, real	1.05	2.06	3.11	3.88	4.00	4.03	4.23	4.93	4.99
Public consumption, real	9.84	12.58	13.36	11.35	10.18	10.02	10.78	10.72	9.66
Real gross investment	5.26	6.64	6.24	3.25	2.17	3.51	4.74	3.57	1.93
Investment in copper	0.50	1.37	−1.45	0.22	0.06	0.09	0.12	0.45	−1.17
Other private investment	0.21	−1.43	−2.79	−4.06	−2.79	−1.78	−1.53	−1.48	−1.86
Total urban employment	1.32	1.89	2.22	2.18	1.96	1.94	2.09	2.29	2.12
Total wage bill	1.08	1.55	2.03	2.17	1.88	1.89	2.08	2.40	1.83
Total operating surplus	2.67	2.72	3.45	3.09	3.00	2.92	3.09	4.37	7.02
Money supply	−0.53	0.19	1.13	2.23	2.80	2.33	2.36	2.51	2.77
GDP price deflator	−0.42	−0.58	−0.48	−0.19	−0.04	−0.26	−0.32	0.30	0.20
Consumer price index	−0.37	−0.41	−0.31	−0.12	−0.02	−0.08	−0.13	−0.10	−0.11
Government current expenditure	9.71	16.01	17.82	13.68	11.46	11.46	12.70	12.25	8.76
Government direct investment	13.38	20.77	20.69	15.97	14.12	15.50	17.11	15.02	11.97
Government current revenue	2.13	2.54	2.58	3.40	3.69	3.47	2.64	4.29	3.93
Imports of goods and services	3.32	4.22	4.62	3.03	2.63	3.74	4.03	3.93	3.05
Exports of goods and services	0.00	−0.01	−0.01	−0.02	−0.02	−0.02	−0.02	−0.02	−0.01
	Differences								
Government net budget deficit (−)	−31.50	−52.40	−64.30	−48.70	−34.10	−38.90	−57.80	−66.20	−41.60
Balance of trade	−12.00	−14.00	−16.00	−12.80	−10.70	−14.00	−21.50	−24.70	−17.30
Balance of current account	−16.60	−17.80	−18.60	−12.30	−8.10	−12.90	−21.30	−22.70	−9.10
Net international reserves	−12.60	−26.30	−40.50	−50.10	−56.50	−66.50	−82.90	−100.50	−107.80

expenditure is able to offset some of the effects of the fall in the price of copper. Real GDP rises throughout the period, with most of the increase coming from manufacturing and services. Urban employment also increases with the offset. The increase in real GDP in the offset solution is able to completely offset the decline in real GDP caused by the copper-price decline in the earlier years, but, toward the end of the solution period, the 10 percent increase in expenditure is not sufficient to offset the GDP decline. However, employment in the offset solution is larger than in the base solution. This is because of the shift from capital-intensive mining to the more labor-intensive manufacturing and services sector.

As in the last section, this offset policy generates deficits in government spending, which have to be financed by domestic and foreign borrowing, and balance-of-payments deficits, which have to be financed from reserves or foreign borrowing. The financing of budget deficits from domestic sources is not inflationary in this solution because its effect on the money supply is offset by deficits in the balance of payments. The same qualification about the feasibility of this offset policy that was made in the previous section applies here also. It depends on the additional domestic and foreign resources the government is able to mobilize. This includes skilled manpower, which is generally in short supply in Zambia.

Stable Growth Rate of Government Spending. A stable growth rate of government spending may be an indirect way of offsetting the effect of copper-price fluctuations. Under this policy, Zambian government spending is not altered in response to short-term changes in copper prices but is fixed and is based on long-run expectations of revenues. This section explores the impact of such a policy on the Zambian economy. To test this policy, we compare a solution in which government expenditures are exogenously set to grow at fixed growth rates with a control solution in which government spending is endogenous and responds to changes in copper prices.

The four endogenous categories of government spending—general, social, and economic services and direct capital expenditures—are exogenized and set to grow at constant growth rates. In addition, government loans to and equity investment in the private sector, an important determinant on noncopper private investment, is exogenized and a constant growth rate is applied to it. The growth rates are chosen so that the growth rate and terminal value of total government expenditures are close to those of the control solution. In addition, the average deficit of government spending in the constant-growth solution should be close and should not exceed that of the control solution. The following growth rates of spending variables meet these conditions: government expenditure on general services (CGGS), 10 percent; government expenditure on social services (CGSS), 10 percent; government expenditure on economic services (CGES), 10 percent; government direct investment expenditure (CGDC), 6 percent; and government loans and investments (GCLIC), 7 percent.

Table 7–5 shows the terminal values, means, and log-trend-adjusted standard deviations of selected variables from the control solution, with endogenous government spending, and the constant-growth government-spending solution. The most notable impact of stabilizing government expenditures is the reduction of the levels of the deflator for GDP, the consumer price index, and its variance. The variance of the deflator for GDP is slightly increased, a counterintuitive result that we shall deal with later.

On the average, there is a shift in real output from services to manufacturing, agriculture, and construction, although, toward the terminal period, this shift was reversed. The primary reason for this is a shift in expenditures from current to capital spending. However, since the assumed growth rates of current expenditures are higher than those of capital expenditures, in line with recent historical experience, the shift away from services was reversed toward the end of the simulation period.

The control solution has a much higher average level of total government spending and higher net budget deficits but a lower terminal value of government expenditures. Also, the levels of real output and employment are higher in the stable-growth solution than in the control solution, although with less government spending on the average. The increase in economic activity came primarily from the manufacturing, agriculture, and construction sectors because of the redistribution of government spending toward investment. This redistribution may be the reason for the superior economic performance under the stable-growth money solution. This is really the heart of the matter. Recent experience of Zambian fiscal policy shows that investment expenditures bear more of the impact of fluctuations of revenues. When revenues fall, investments fall, but when revenues rise, resource constraints prevent the increase in investment to compensate for past reductions. Thus, the level of investment may be lower than the government will rationally desire. If a stable growth rate of capital and current expenditures keeps investment expenditures in line with desired investment, it will be good for economic growth.

The variance of gross domestic output, adjusted for trend, in the stable-growth solutions exceeds that of the control solution. The trend-adjusted variances of consumption and employment are also higher for the stable-growth solutions, while those of gross investment, consumer price index, government budget deficit, and money supply are lower. The expectation that stabilizing the growth rates of government expenditure would stabilize the major real variables has not been realized from the results of the simulation tests. We shall provide some explanation for this seemingly implausible result at the conclusion of the next section.

Interdependence of Fiscal and Monetary Policy

We have just examined separately the impact that stable money supply and fiscal expenditures had on the performance of the Zambian economy. Monetary and

Table 7-5

Comparison of an Endogenous Government-Spending Solution with an Exogenous Constant-Growth Government-Spending Solution

	Endogenous Government Spending			*Exogenous Government Spending*		
Variables	*Terminal Value*	*Mean Value*	*Standard Deviation*	*Terminal Value*	*Mean Value*	*Standard Deviation*
Real GDP	1452.5	1071.4	0.0483	1521.9	1074.6	0.0536
Agriculture	190.2	148.8	0.0224	190.9	149.8	0.0244
Mining and quarrying	176.9	184.9	0.1670	176.9	184.9	0.1669
Manufacturing	184.4	135.1	0.0804	186.5	137.8	0.0860
Services	726.0	467.6	0.0382	788.7	464.2	0.0350
Private consumption, real	535.0	435.5	0.1071	568.9	437.4	0.1167
Public consumption, real	345.0	245.5	0.0919	437.7	240.5	0.1117
Real gross investment	402.7	309.2	0.1437	402.6	324.5	0.1086
Government recurrent revenues	2307.1	1087.8	0.1645	2397.2	1090.5	0.1748
Government capital expenditures	480.7	240.6	0.1419	477.8	270.9	0.0000
Government recurrent expenditures	1703.1	849.3	0.0961	1724.0	751.2	0.0413
Government loans and investments	245.5	130.5	0.2191	297.7	156.1	0.0000
Government net budget deficit (–)[a]	–104.3	–119.9	155.3	–84.5	–75.0	110.8
Total urban employment	508.8	416.3	0.0255	533.9	417.1	0.0260
Money supply	2955.6	1367.2	0.1040	2673.3	1256.3	0.0969
Deflator for GDP	4.335	2.835	0.0867	4.275	2.822	0.0881
Consumer price index	199.4	138.0	0.01104	195.8	136.3	0.1096

[a]The standard deviation for this variable is not trend adjusted but is calculated from levels.

fiscal policies are not independent in the Zambian context. Theoretically, it is possible to change government spending and leave money supply unchanged, but in the present stage of development of financial institutions in Zambia, that is hardly a routine operation. An increase in Zambian government expenditure can be financed by (1) drawing on government accounts in the Central Bank and the commercial banks; (2) borrowing from abroad; (3) government borrowing from the Central Bank, the commercial banks, and the nonbank public; and (4) taxation.

When the government borrows from the Zambian Central Bank or draws on its accounts with the bank and spends the money, the money supply is increased. Borrowing from domestic commercial banks by the sale of treasury bills or other debt instruments also leads to monetary expansion, since these debt instruments can be applied to meet legal reserve requirements. Drawing down accounts from the commercial banks should not lead to monetary expansion unless the banks maintain excess reserves, which is usually the case in Zambia. Financing a deficit by borrowing from the nonbank public does not lead to monetary expansion, but this is a minor market for selling government debt. According to government finance statistics, the major domestic sources of government borrowing in the past have been the Central Bank (54 percent), the commercial banks (36 percent), and the nonbank public (10 percent).[6] Foreign loans and grants have in the past financed about 29 percent of government deficits. Foreign borrowing does not lead to monetary expansion if the proceeds are used abroad. Taxation is a possible but not feasible method of meeting government short-term financial needs because of collection costs and its effect on the level of uncertainty in the economic environment.

In the simulation reported in the last section and in table 7–5, monetary policy was accommodating to the fiscal policy simulated. The results indicated that stabilizing government expenditures reduced the average level and trend-adjusted variance of the money supply. However, this result may not be robust and may depend on the pattern and variance of copper prices. Money supply responds to government deficit or surplus and to changes in net foreign assets. A stable expenditure policy may therefore destabilize the money supply and consequently the price level. To be able to control the money supply effectively, more of the burden of financing the government deficit should be borne by individuals and institutions other than the Central Bank and commercial banks. This will require changes in financial institutions and operations that will make government short-term debt available and attractive to the nonbank public.

We examine the effects of insulating both the government expenditures and the money supply from the effect of copper-price fluctuations by adopting constant fiscal and monetary growth rules. Thus, we obtain a solution that incorporates a 13 percent exogenous growth rate of money supply and the constant growth rates of government expenditures outlined in the last section. Table 7–6 presents the terminal values, means, and trend-adjusted standard deviations of selected variables. We compare this table with results in tables 7–2 and 7–5, which have the

Table 7-6
Impact of Exogenous Constant Growth Rates of Government Spending and Money Supply

Variables	*Terminal Values*	*Mean Values*	*Standard Deviation*
Real GDP	1537.1	1076.0	0.0562
Agriculture	191.4	149.6	0.0240
Mining and quarrying	176.8	185.0	0.1667
Manufacturing	187.2	137.3	0.0843
Services	788.0	464.8	0.0356
Private consumption, real	588.5	436.3	0.1260
Public consumption, real	436.8	241.3	0.1131
Real gross investment	452.7	327.8	0.1274
Government recurrent revenues	2443.7	1091.4	0.1784
Government capital expenditures	477.8	270.9	0.0000
Government recurrent expenditures	1724.0	751.2	0.0413
Government loans and investments	297.7	156.1	0.0000
Government net budget deficit or surplus[a]	–37.9	–74.1	111.14
Total urban employment	534.5	417.3	0.0270
Money supply	3747.5	1331.1	0.0000
Deflator for GDP	4.296	2.815	0.0883
Consumer price index	202.2	135.9	0.1062

[a]Standard deviation for this variable is calculated from levels.

control solution and separate solutions for constant monetary and fiscal growth rates.

The combined stable-monetary and fiscal-growth-rates policy is effective in reducing the average level and variance of the consumer price index. It also gives the same counterintuitive result, that stabilizing fiscal spending or money supply does not lead to stability of real variables.

The answer to this unexpected result may lie in the behavior of domestic prices in response to copper-price fluctuations. In the absence of any countervailing policy, domestic prices will move with copper prices, tending to rise when copper prices rise, and the reverse. To simplify this explanation, we introduce some notation. Let the nominal shock on a variable (for example, investment) caused by a change in copper prices by z. Then the real shock y is given by $y = z/p$, where p is the domestic deflator for investment. With no countervailing policy, z and p would move up and down together, thus tending to reduce the variability of y. With a policy that stabilizes p or cuts off the link between z and p, when z changes, p does not necessarily change, tending to increase the variability of y.

Mathematically, we can write that as $\log y = \log z - \log p$ and $\text{var}(\log y) = \text{var}(\log z) + \text{var}(\log p) - 2\,\text{cov}(\log z, \log p)$. This relationship shows that the variance of $\log y$ will fall, ceteris paribus, if the covariance between $\log z$ and $\log p$ rises. For the variance of $\log y$ to rise when the variance of $\log p$ changes implies that the covariance term falls more than the variance of $\log p$. The only policy that will be effective in reducing the variance of log y is one that reduces the variance of log p more than the covariance of log p and log z. The results of the simulations indicate that stabilizing p through the stabilization of money supply or government expenditures is not effective for reducing the variance of log p more than the covariance of log p and log z. The analysis implies that a certain amount of fluctuation in p, which increases its covariance with the nominal shock of copper prices, is necessary for a real stability.

It was also found that stabilizing the money supply or government spending does not reduce the variability of the deflator for GDP. This deflator is a weighted average of the price indexes of the various supply sectors, of which mining and agricultural indexes are determined by factors that are exogenous in the model. The variance of deflator for GDP may rise as a result of the increase in the variances of the weights or the interactions of the several covariance factors.

Commercial Policies in the Copper Industry

The Zambian government owns a controlling interest in the two producing mining companies.[7] This put it in a position of great influence in the determination of the objectives and operating policies of the companies.[8] The minority partners are concerned with maximizing the present value of their investment, which may be achieved by maximizing profits in each accounting period. The government has broader objectives, including maximization of mineral revenues, employment, foreign exchange, and economic growth and offsetting the impact of copper-price fluctuations. All these objectives cannot be met by a policy of profit maximization, at least in the short run. The government has always stressed the need for increasing levels of output because it believed that higher levels of output would generate more foreign exchange and mineral revenues necessary for Zambian development in addition to creating more employment.

In the earlier years following independence, the structure of costs and prices was such that increasing levels of output would most probably increase profits. However, this would have demanded new investment, which the mining companies were reluctant to undertake because of their concern with taxation and future copper-price outlook and their fear of possible expropriation. There was thus a disparity between the desires of the government and what the companies were willing to do. Their disagreement on investment might have been that the two sides were using different implicit or explicit discount rates in evaluating investment proposals. This is principally the result of different risk perceptions, the mining companies having to discount for political and expropriation risks in addition to other operating risks. This problem would still be relevant after the takeover, although one would expect that the role of expropriation risk would diminish.

The government would be interested in maximizing output, since it could produce higher revenues, employment, and foreign exchange, at least in the short run. Assuming that changes in Zambian production have no impact on world copper prices, then, for a given level of prices, foreign exchange will be maximized by a higher level of output so long as the marginal cost of imported input requirements is less than the price. Imported inputs include materials and labor and payments for foreign capital (dividends). We assume that this condition is met in the Zambian case for the levels of prices and production that we consider, so that, in the short run, output maximization implies the maximization of foreign-exchange earnings. This policy alone, however, may not maximize foreign exchange in the long run, since an inventory policy that concentrates sales when prices are high, and the reverse, may dominate a policy of simply selling all finished output. Our present model is not suited for an investigation of the implications of such a policy, so we must defer it for later work.

The differing objectives of the government and the minority partners appear to be recognized in the takeover agreement, which stressed the need for the "optimization of production and profit" and a government undertaking to permit the operating companies to "conduct their business on a commercial basis."[9] The minority partners interpreted the agreement as implying only profit maximization;[10] the government position was not clearly articulated, but its objective to maximize production and employment in the mines was the major motivation for the takeover. It could be that the government assent to what appears to be a policy of profit maximization was tactical rather than strategic; thus, an explicit policy of production maximization could be adopted in the long run.

We examine the long-run effects of a policy of profit maximization (PMP) and a policy of copper-output maximization (OMP). The two policies are distinguished by the motivation behind output decisions. In the profit-maximization case, copper output (supply) is a function of real market prices of copper and is constrained by available production capacity, while in the output-maximizing case, output is equal to capacity output.[11] In both cases, output depends on investment decisions made in the past, since production capacity is represented by capital stock existing at the beginning of the period. Capacity output of copper is the upper ceiling of production under PMP.

Investment decisions in a profit-maximizing framework are determined by anticipation of market prices, costs of production, which determine factor proportions, capacity utilization, and costs of financing. In our model, investment in the copper industry is largely determined by price expectations, represented by past prices, and, to a smaller extent, lagged profits. This formulation falls into the profit-maximizing framework. In the output-maximization framework, past investments completely determine future output, and so investment should depend on desired levels of future output. Even in this case, the level of market prices ought to be an important factor in investment decisions, since it influences the ability to finance new investment. In order to examine the role of investment, we shall consider cases where investment is determined exogenously (that is, independent of market prices) and solve the model for different levels of investment.

In the simulations that follow, the same decision model is assumed for employment of inputs of production in both the output-maximization and the profit-maximization scenarios. Employment of Zambians depends on capacity utilization, capital-output ratios, and the level of output, while expatriate employment is a function of output and expatriate wage levels. An important variable in any solution under output maximization is capacity utilization. It is determined endogenously in the PMP solution but will be exogenized for the OMP solutions. The historical series was determined by the peak-to-peak method,[12] using the years 1965 and 1969 as the recent peaks. We shall investigate the effects of changes in capacity utilization under output maximization. Output is maximized when capacity is fully utilized, but it would be unrealistic to expect 100 percent utilization in all periods. Output is affected by other factors besides deliberate decisions not to go at full capacity, such as shortages of required inputs, breakdown of equipment, and industrial disputes and accidents.

Employment in the copper sector will be maximized in the short run by an output-maximization policy, although higher utilization of capacity increases labor productivity but the marginal input of labor is always postive. In the long run, the impact is not clear; it will depend on such factors as profitability and its effect on future output through investment and the effect of changes in cost on investment and factor proportions. Output maximization will also maximize total employment in the economy in the short run but not necessarily in the long run. Employment in noncopper sectors will be affected to the extent that output maximization affects government revenues, which has a direct impact on investment, output, and employment in the noncopper sectors.

Simulation Tests

The model is simulated under assumptions of profit-maximization policy (PMP) and output-maximization policy (OMP). Copper prices used in the simulations are assumed to have a secular-trend growth rate of 5.5 percent, fluctuating around this trend by a standard deviation equal to that of the historical series. The solution generated under the profit-maximization assumption, with endogenous investment and capacity utilization, is used as a control solution, and the other solutions are compared to it.

We first examine the effect of OMP under different rates of capacity utilization. Table 7–7 shows the percentage differences for selected variables from solutions under OMP with endogenous investment, for exogenous capacity-utilization rates of 100 percent and 95 percent. The results are presented for even years up to 1976 and from there for every fourth year and the terminal year. Higher levels of utilization produced higher production, employment, wages, and profits in the copper industry. Expatriate employment appears very sensitive to increases in production, the elasticity of expatriate employment with respect to output

Table 7-7
Effect of Capacity Utilization under a Policy of Output Maximization

	Percentage Differences between Solutions with 100 Percent and 95 Percent Capacity Utilizations								
Variables	*1968*	*1970*	*1972*	*1974*	*1976*	*1980*	*1984*	*1988*	*1990*
Copper sector:									
Production (tonnes)	9.50	19.90	16.80	28.30	21.30	25.60	24.40	22.90	22.00
Employment: Zambians	4.50	9.10	5.40	14.30	8.40	12.00	11.10	10.00	9.30
Employment: expatriates	2.40	9.40	14.60	24.30	29.40	44.70	65.80	80.90	106.40
Average wages: Zambians	5.00	11.70	14.80	17.00	15.90	16.70	16.60	17.30	17.30
Average wages: expatriates	2.80	6.40	6.70	6.30	2.30	–0.50	–2.90	–3.90	–4.50
Industry profits	7.90	18.70	16.40	21.10	18.00	20.40	20.40	18.00	17.70
Total tax liability	9.50	19.10	20.00	25.00	19.30	23.00	22.30	20.20	19.60
Real investment	–1.30	–0.10	–2.00	–1.90	–1.30	–0.40	–1.10	–1.10	–1.30
Value added by industry, real	7.50	19.60	15.30	23.50	26.30	28.70	26.30	24.60	23.40
Economywide:									
Real GDP	2.10	5.50	5.10	6.80	7.90	6.60	5.80	4.70	4.20
Agriculture	–0.03	0.04	0.30	0.60	0.70	0.90	1.00	1.10	1.10
Manufacturing	–0.02	1.60	4.60	5.70	6.10	6.90	7.20	7.00	6.70
Services	–0.25	1.10	0.90	0.70	0.20	0.70	0.40	0.30	0.20
Private consumption	2.30	6.40	5.20	5.40	2.60	2.20	0.33	–1.20	–1.70
Public consumption	–1.10	1.90	1.10	0.70	0.40	0.70	0.20	–0.02	–0.40
Gross real investment	0.30	7.60	8.20	13.00	9.70	11.20	10.60	10.10	9.40
Total urban employment	0.80	2.70	3.20	5.00	4.40	5.20	5.00	4.60	4.30
Total wage bill	4.20	12.10	13.80	17.80	13.70	14.70	14.40	14.60	14.40
Operating surplus	6.60	17.70	12.00	10.60	3.70	5.10	8.00	3.80	3.80
Money supply	7.20	19.80	25.00	30.90	29.20	27.30	25.90	24.00	22.70
Consumer price index	1.40	3.40	4.20	4.60	3.90	3.60	3.70	3.90	4.00
GDP price deflator	1.90	4.60	4.60	4.70	1.80	3.50	4.20	4.50	4.60
Government current revenue	7.10	16.20	15.90	17.40	12.90	12.00	12.30	11.00	10.80
Government current expenditure	1.10	9.20	10.90	10.80	6.10	7.40	6.50	6.30	5.80
Government capital expenditure	1.50	12.50	15.60	18.90	12.50	10.50	10.70	10.50	10.30
Net international reserves	20.10	46.90	55.20	99.60	77.10	62.70	61.30	62.90	62.80

reaching 2.0 by 1984 and exceeding that value in subsequent years. The employment of Zambians is not so sensitive, reaching a maximum elasticity of about 0.5 in 1974. This is partly because higher utilization rates tend to increase the productivity of local labor but are not significant for expatriate labor. Average annual wages of Zambians in copper are higher for the higher utilization rate, as a consequence of higher productivity and industry profits. Expatriate wages are higher initially but decline from 1980, resulting from a decline in the productivity of expatriate labor.

Real gross domestic product is higher for the 100 percent utilization, the mining and manufacturing sectors contributing most of the increase. Manufacturing value added is higher, since real gross investment by the noncopper sector is increased. Mineral revenues of the government are higher at 100 percent utilization, thus generating more funds for government direct investment and government loans to and investment in the private sector. At the higher utilization rate, the increase in net foreign assets causes the money supply to rise, which induces higher aggregate domestic price levels. The higher price level is partly responsible for the fall in real personal and public consumption at the higher utilization rate toward the end of the simulation period. The other influence on domestic consumption is the change in the proportion of wage income spent domestically. At the higher utilization rate, the proportion of wage income due to expatriates is higher, and hence the proportion of wage income spent domestically is lower toward the end of the simulation period.

Table 7–8 shows comparisons of solution values of selected variables under PMP and OMP with 100 percent capacity utilization. The solution under OMP gives higher copper-industry output, employment, profits, and value added than does the profit-maximizing solution. The increase in activity in the copper sector induces higher levels of activity in other sectors, especially manufacturing and construction; thus, overall real gross domestic product and employment are higher. With increases in copper production, exports and foreign-exchange earnings are higher. The level of international reserves is higher for the OMP solutions, which also have higher domestic price levels and slightly lower investment in the copper sector. The increased price level reduces the real value of nominal investment in the copper sector. This investment is largely determined by exogenous copper prices, which have the same values in all the solutions.

Changes in copper-industry investment will affect the results of these solutions. There does not seem to be much to be gained by solving the model under different levels of investments, as we can infer the results from the simulations already performed. Increasing the levels of real investment exogenously above the levels in the control solution will have an insignificant effect on activity in the copper sector under PMP, since in this case copper output is largely determined by past and present prices of copper. It will have slight macroeconometric effects, depending on how the new investment is financed. If it is financed by foreign loans, this capital inflow will increase the domestic money supply and hence the price level. These effects will be mitigated by an increase in imports caused by the

Table 7-8
Comparison of Economic Performance under PMP and OMP, with Full Utilization of Capacity

	Percentage Differences								
Variables	*1968*	*1970*	*1972*	*1974*	*1976*	*1980*	*1984*	*1988*	*1990*
Copper sector:									
Production (tonnes)	9.20	15.20	20.20	40.50	47.90	42.90	29.70	16.50	9.50
Employment: Zambians	1.20	4.30	7.00	14.70	22.40	18.80	14.50	8.80	5.90
Employment: expatriates	2.30	7.20	13.00	28.40	48.00	85.80	119.70	97.90	89.90
Average wages: Zambians	7.30	10.20	13.90	29.20	29.60	29.70	20.00	11.70	6.90
Average wages: expatriates	2.90	4.90	6.80	10.00	8.10	–0.80	–5.50	–6.30	–6.10
Industry profits	7.60	14.60	20.60	30.20	41.80	32.50	23.50	11.20	5.90
Total tax liability	9.20	15.00	25.30	36.30	45.80	37.10	25.70	12.50	6.40
Real investment	–1.50	–0.70	–2.00	–2.70	–2.10	–0.90	–1.50	–1.10	–1.10
Capacity utilization	9.20	8.60	8.70	10.00	4.80	8.80	4.40	2.11	0.20
Value added by the industry	7.20	14.90	18.40	33.10	62.50	48.90	32.10	17.60	10.00
Economywide:									
Real GDP	1.90	4.10	5.50	8.50	14.50	9.90	7.30	4.30	2.90
Agriculture	–0.05	0.03	0.20	0.40	0.70	1.00	1.40	1.50	1.50
Manufacturing	–0.04	1.40	3.50	5.20	6.10	9.90	10.30	9.00	7.70
Services	–0.20	0.70	0.50	0.90	0.40	1.00	0.50	0.20	0.02
Private consumption	2.20	4.80	6.20	9.60	9.10	4.20	–1.30	–4.70	–5.90
Public consumption	–1.20	0.80	0.30	0.60	–0.80	0.90	0.20	–0.06	–0.40
Gross real investment	0.02	4.70	6.30	15.30	15.40	17.90	14.70	10.20	7.30
Total urban employment	0.50	1.80	3.00	5.80	7.80	8.30	6.80	4.70	3.40
Total wage bill	4.60	9.30	12.50	24.30	24.30	24.50	19.20	13.40	9.70
Operating surplus	6.10	13.50	15.20	14.10	10.90	7.50	8.70	0.50	–1.70
Money supply	6.90	14.60	21.10	35.60	48.20	47.40	38.30	25.10	18.00
Consumer price index	1.60	2.70	3.60	5.90	6.10	6.00	5.10	4.00	3.10
GDP price deflator	2.10	3.70	4.60	6.60	2.30	5.80	5.50	4.10	3.00
Government current revenue	7.20	12.60	17.30	24.10	23.00	19.40	15.50	9.50	6.50
Government current expenditure	1.10	6.20	8.30	14.00	10.10	11.80	8.60	6.00	4.20
Government capital expenditure	1.50	8.60	12.70	25.00	21.40	20.30	15.30	10.30	7.20
Net international reserves	19.40	32.90	45.50	122.20	157.00	131.50	103.30	70.10	51.60

new investment. If the new investment is financed by a loan from the government without a corresponding reduction in government spending elsewhere, the money supply and the price level will rise. An exogenous decrease in investment under PMP may decrease the level of activity in the copper industry to the extent that desired levels of output dictated by profit maximization are constrained by the availability of production capacity.

Changes in the level of investment in the copper sector will affect activity in that sector under OMP. An increase in the level of investment above that generated endogenously will increase copper-production capacity and therefore output of copper. This will have induced effects on the rest of the economy, as before. The method of financing this additional investment will also have an impact on the money supply and the price level, as was just outlined.

Economic Stability under OMP

This section examines whether a policy of output maximization contributes to economic stability. As usual, we calculate the log-trends-adjusted standard deviations of major economic variables as the measure of stability. These are shown in table 7–9 for two solutions, one based on the profit-maximization policy (PMP)

Table 7-9
Comparison of Log-Trend-Adjusted Standard Deviations under PMP and OMP for Selected Variables

	Standard Deviation	
Variables	*PMP*	*OMP*
Real GDP	0.0483	0.0422
Agriculture	0.0224	0.0225
Mining and quarrying	0.1670	0.1193
Manufacturing	0.0804	0.0919
Services	0.0382	0.0393
Private consumption, real	0.1071	0.0782
Public consumption, real	0.0919	0.0930
Real gross investment	0.1437	0.1150
Total urban employment	0.0255	0.0313
Total wage bill	0.1318	0.1816
Money supply	0.1040	0.1924
GDP price deflator	0.0867	0.0888
Consumer price index	0.1104	0.1225
Government current revenues	0.1645	0.1317
Government mineral revenues	0.4972	0.3892
Government current expenditures	0.0961	0.1146
Government capital expenditures	0.1419	0.1022
Government budget deficit or surplus[a]	115.34	129.97
Copper-industry profits	0.4484	0.3688
Net international reserves	0.1607	0.2330

[a]Standard deviation for this variable is calculated from levels.

and the other on the output-maximization policy (OMP), with full-capacity utilization. The results indicate that OMP reduces the variability of real gross domestic product, private consumption, and real gross investment. There is a substantial reduction in the variability of the value added from the copper industry. OMP does not stabilize the price level; the trend-adjusted variance of the price level is higher than in the PMP solution. This is a result of the increase in the variance of the money supply under OMP. The variability of the copper-industry profits is reduced while that of the net international reserves increases under OMP. We can conclude that a policy of output maximization stabilizes overall real economic activity but increases the variance of the price level. However, the variance of the price level could be reduced by stabilizing the growth of money supply or government expenditure, as we showed in earlier sections of this chapter.

Impact of Changes in Ore Grade

An important determinant of mining operations is the grade of ore mined. In the previous simulations, we have assumed the grade to be independent of the type of operating policy adopted. However, insofar as the operating policy (profit maximization or output maximization) affects the cumulative output or the output in each accounting period, the grade of ore may be dependent on operating policy. As a nonrenewable resource, copper deposits in Zambia deplete at the rate that they are taken out of the ground.

The grade of ore mined may be affected by the rate of depletion in two ways. The high-grade deposits among known reserves tend to be worked first, so that, in the absence of further discoveries, the grade declines as more ores are raised from the mines. In a profit-maximizing framework, the high-grade ores get worked intensely, while lower-grade ores are worked if the price is right. When the copper price rises, output may be increased by working those lower-grade deposits, whose marginal costs are lower than the price. When the price falls, those mines whose marginal costs are higher than the price are closed down. In an output-maximizing framework, all producing mines are worked, irrespective of price. All things being equal, we should expect the grade of ore to be lower in an OMP framework than in a PMP situation, if the profit maximizer follows the strategy of closing low-grade mines first. Again, all things being equal, we should expect an inverse relationship between the grade of ore mined and the cumulative depletion of reserves and the level of output in each accounting period.

The kind of operating policy adopted may affect the grade of ore in different ways. A profit-maximization policy may attract more funds for exploration, especially from foreign sources. More intensive exploration could lead to the discovery of new deposits, which may have high grades. High depletion rates, which result from output maximization, may induce the government and industry to explore more for new deposits. However, these qualifications will not be relevant in the limit, when all deposits are known.

To incorporate the dependence of ore grade on the rate of depletion of deposits in our analysis, we have to obtain a relationship between the ore grade, the cumulative copper output, and the output in each period. This relationship is obtained by regressing the average ore grade (VOGRADE) against the period output (CUQCZM) and the cumulative output (CUMQ). CUQCZM was not significant when used with either contemporaneous or lagged CUMQ. Only CUMQ entered significantly with the expected negative sign. The estimated relationship[13] for the 1953–1977 period of fit was

$$\underset{}{\text{VOGRADE}} = \underset{(8.94)}{8.6491} - \underset{(-5.7)}{0.625560} \log(\text{CUMQ})$$

$$\bar{R}^{2} = 0.57 \qquad SEE = 0.37 \qquad DW = 1.05 \tag{7.1}$$

where CUMQ was cumulative from 1949 from CUQCZM.

The grade of ore (VOGRADE) is normally exogenous in the model. To perform a comparative analysis of profit-maximization and output-maximization policies incorporating grade effects, VOGRADE was endogenized outside the model. The values of grade under PMP and OMP are calculated using equation 7.1 and values of CUQCZM from the PMP and OMP solutions compared in table 7–8. These two solutions have the same values of grade, and the OMP solution assumes full capacity utilization. We therefore obtain two series for ore grade, which are different because the cumulative outputs are different. By 1990, the values of grade differ by as much as 8 percent. We incorporate the two new series to obtain new solutions for PMP and OMP with full capacity utilization.

The two solutions incorporating grade effects are compared in table 7–10. This table should be compared with table 7–8, which has the same solutions but assumes same values of grade. The differences between the PMP and OMP solutions in table 7–10 are much smaller. For example, by 1976, the percentage difference in copper production between the solutions with the same grade is 47.9, compared to 32.8 in table 7–10. By 1990, output, profits, tax liability, and value added by the industry are higher for the PMP solution than for the OMP solution. Thus, when depletion effects are considered, through their impact on the grade of ore, output maximization is still more stimulative to the copper sector and the macroeconomy for the period examined, but it may not be so in the very long run.

The foregoing simulations indicate that an output-maximizing policy dominates a profit-maximizing policy for both the copper industry and the economy, even when the effects of depletion on the grade of ore are taken into consideration. However, this result has to be qualified. The simulations assume that the structure of costs remains the same as in the estimation period, that is, average costs rise very slowly with increases in production, such that capacity output may be attained without the marginal cost exceeding the price of copper. The mining companies

Table 7-10

Comparison of Economic Performance under PMP and OMP, with Full Utilization of Capacity and Endogenous Ore Grade

	Percentage Differences								
Variables	*1968*	*1970*	*1972*	*1974*	*1976*	*1980*	*1984*	*1988*	*1990*
Copper sector:									
Production (tonnes)	7.10	10.80	7.60	26.80	32.80	26.50	12.50	1.90	–0.07
Employment: Zambians	0.90	3.10	3.70	9.20	16.10	11.70	6.70	1.30	–0.05
Employment: expatriates	1.80	5.20	6.70	17.40	32.10	57.00	69.50	42.90	36.50
Average wages: Zambians	5.70	7.50	5.90	20.10	20.50	19.20	9.00	2.00	0.60
Average wages: expatriates	2.30	3.70	3.10	7.00	6.20	–1.00	–4.60	–4.40	–3.50
Industry profits	5.90	10.40	7.40	20.30	28.80	19.80	8.70	–0.30	–1.40
Total tax liability	7.10	10.60	9.10	24.60	31.50	22.60	9.70	–0.40	–1.50
Real investment	–1.20	–0.50	–1.00	–2.00	–1.50	–0.80	–1.00	–0.60	–0.60
Capacity utilization	7.10	5.90	1.50	7.00	2.10	5.80	1.10	0.00	0.00
Grade of ore mined	–0.20	–0.70	–1.30	–2.50	–3.90	–6.10	–7.50	–8.00	–8.00
Value added by industry	5.60	10.60	6.90	21.80	42.90	30.20	13.50	2.10	–0.08
Economywide:									
Real GDP	1.50	2.90	2.20	5.50	9.70	6.20	3.50	1.20	0.60
Agriculture	–0.04	0.04	0.02	0.02	0.04	0.60	0.90	0.90	0.80
Manufacturing	–0.03	1.10	2.50	3.40	3.60	6.60	6.40	4.70	3.60
Services	–0.20	0.50	0.30	0.60	0.20	0.60	0.20	–0.05	–0.10
Private consumption	1.70	3.40	1.90	6.30	6.80	2.60	–2.30	–4.20	–3.80
Public consumption	–1.00	0.70	0.40	0.10	–0.60	0.50	0.04	–0.20	–0.30
Gross real investment	0.00	3.70	3.60	9.60	10.40	11.60	8.00	3.30	1.60
Total urban employment	0.40	1.40	1.60	3.70	5.30	5.40	3.70	1.60	0.90
Total wage bill	3.60	6.90	6.20	15.80	16.60	15.88	9.80	4.20	2.50
Operating surplus	4.70	9.50	5.20	9.60	6.90	4.10	2.30	–2.90	–3.50
Consumer price index	1.30	2.10	2.10	3.90	4.40	4.20	3.00	1.50	0.90
GDP price deflator	1.60	2.80	2.10	4.50	1.70	4.00	2.90	1.30	0.70
Government current revenue	5.60	9.10	7.10	16.00	15.70	12.50	7.50	2.50	1.20
Government current expenditure	0.90	4.80	4.80	8.70	7.10	7.70	4.60	1.80	0.90
Government capital expenditure	1.10	6.50	6.90	15.90	15.10	13.20	8.00	3.00	1.50
Net international reserves	15.00	25.00	27.10	83.70	112.60	87.20	57.60	27.90	17.60

may not have been profit maximizing in the sense that the scale of their operations was not optimal. They may have been constrained by political risk from increasing the scale of their operations; the additional expected return would not compensate for the increased risk of their portfolio of holdings by additional commitments to Zambia. Another factor that this analysis may not have taken into consideration fully is the effect of increased output in each period on the grade of ore. A mine-by-mine analysis would indicate those mines that are not profitable at given prices; they may then be shut down while the profitable ones work at full capacity. However, an aggregated analysis would make a decision to reduce output on the basis of average marginal costs across mines, which may be less than price, while the average marginal cost of some mines actually exceeds the price. This analysis also assumes that reductions and increases in output occur across mines when such decisions are implemented.

With these qualifications, the result that increased scale of operations by output maximization increases profit is very tenuous, and its effect on government revenue and spending decisions is equally tenuous. Thus, the merits of output maximization lie in its effect on increasing foreign exchange and employment. Output maximization, coupled with an inventory policy that concentrates sales when the prices are high, and the reverse, may be used to maximize foreign-exchange earnings from copper, although the costs of such an inventory policy may be prohibitive.

Agricultural Policies

The development of a strong and productive agricultural sector is one of the major goals of the Zambian government. Such a development would enable the nation to achieve several objectives: (1) becoming self-reliant in food production; (2) correcting regional economic imbalance and narrowing the rural-urban income gap; (3) diversifying economic activity and exports away from copper; (4) satisfying the need for agricultural-commodity inputs to domestic processing industries and reducing the need to import these commodities; and (5) contributing to overall growth.

Between 1966 and 1976, real agricultural value added grew at an average annual rate of 2.7 percent. This compares with an average annual growth rate of 2.4 percent for real gross domestic product, 7.7 percent for the manufacturing sector, 6.6 percent for services, and 3.1 percent for total population. The proportion of real gross domestic product from agriculture has remained at about 14 percent during this period, as shown in the top line in figure 7–1. The middle line shows the behavior of agricultural income as a proportion of nominal GDP, and the bottom line is the proportion of real GDP from rural agriculture. Agricultural income as a proportion of GDP has fluctuated considerably, primarily as a result of variations in nominal GDP by the fluctuations of copper prices. For example, in

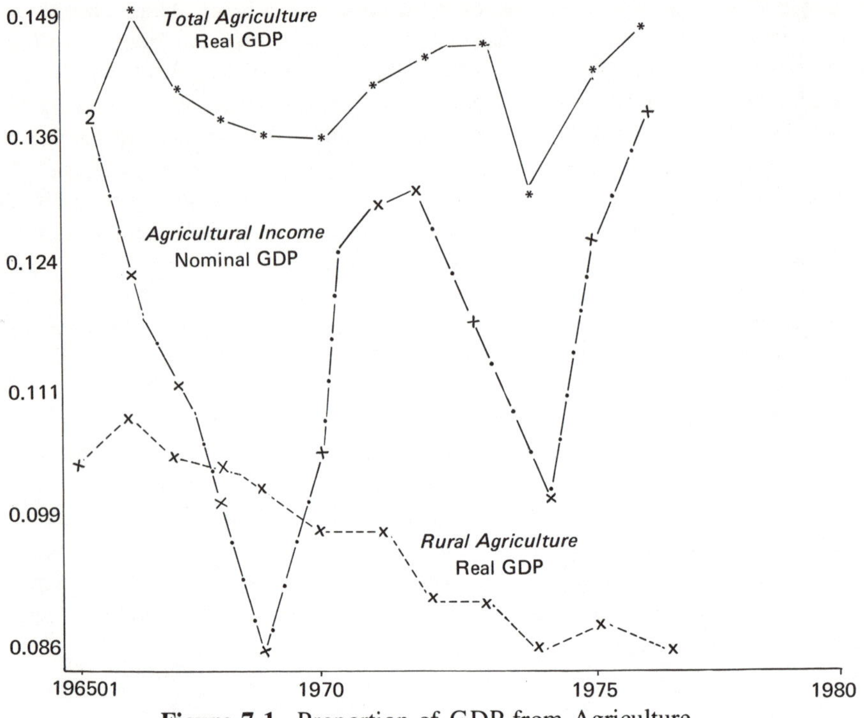

Figure 7-1. Proportion of GDP from Agriculture

1969 and 1974, copper prices were at cyclical highs and the proportion of mining incomes was high, thus reducing the proportional contribution of other sectors. Nevertheless, the important point is that agricultural income as a proportion of GDP is less than its proportion in real terms of all years.

The graph in figure 7–2 compares the performance of the rural and commercial agricultural sectors. The top lines are the proportions of real and nominal agricultural value added from the rural sectors; the bottom line is the proportion of real agricultural value added from the commercial sector. This graph shows that the proportion of output from the rural sector is declining. The commercial sector grew by 7.7 percent between 1965 and 1976, while the growth rate for rural agriculture was 0.9 percent. The tendency of the declining proportion of rural agriculture is caused by several factors. There has been a greater emphasis on commercial agriculture through more investment by private and state enterprises and through the entry of farmers from rural, labor-intensive farming to the capital-intensive commercial sector. Productivity in rural agriculture may be declining , while that in commercial agriculture is increasing through the use of more capital-intensive methods. Migration from rural to urban areas has robbed the rural sector of the more productive manpower.

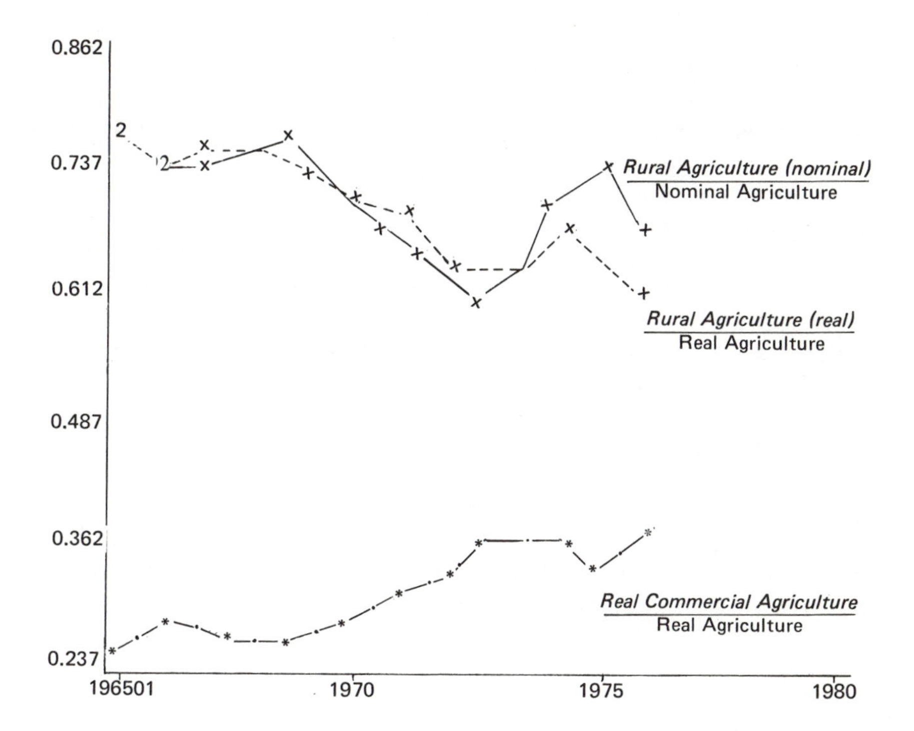

Figure 7-2. Proportion of Agricultural Value Added from Rural and Commercial Sectors

Since 1965, imports of food have increased. Food imports (excluding beverages and tobacco) as a proportion of private consumption have fluctuated between 9 percent in 1971 and 3 percent in 1976. However, there is an upward trend in the proportion of food imports in private consumption. Zambia has had to import substantial quantities of maize, its staple and its major agricultural product, to meet domestic needs. Exports of the two major agricultural export commodities—maize and tobacco—have declined. Maize exports declined from 2 million kwacha in 1965 to 0.5 million kwacha in 1976, while tobacco declined from 5 million kwacha to about 2 million kwacha.

The foregoing analysis indicates that the agricultural sector is far from meeting the objectives that have been set for it. Rural income from agriculture is declining as a proportion of total incomes, and the rural-urban terms of trade appear to be biased against the rural sector. Regional imbalance is not corrected, since regional imbalance and rural-urban disparity are closely related. Imports of agricultural commodities are on an increasing trend, while exports are declining, so that the trend is away from self-reliance in food and diversification from copper.

Thus, if the stated objectives are to be met, there is need for action to reverse the trends that are apparent in past performance.

The problems facing Zambian agriculture are many, and the nature of these problems may be economic, institutional, political, social, managerial, and climatic. Zambia is a sparsely populated country, and there is no shortage of good agricultural land. Commercial agriculture is concentrated in areas with good transportation networks and very rich soil, while rural agriculture is scattered all over the country. Managerial skills are in short supply, and most of the available skills are employed in the modern nonagricultural sector. The political power is asymmetrically distributed in favor of the urban nonagricultural population, and the agricultural pricing policy of the government seems to favor the urban dwellers. The financial institutions are not geared to serve the rural areas, most of their operations being concentrated in the urban areas. The rural farmers do not usually have the kind of collateral that is desired by a conservative banking system, nor do they possess the skills necessary to communicate effectively with such institutions. Without access to credit, new investment must come from savings, which limits the ability of the sector to expand. We have to abstract from some of these issues and problems, although they are of great importance. Our major objective is to contribute a quantitative dimension to the economic issues affecting the sector, and we now turn to that.

Agricultural Pricing

The Zambian government controls the pricing and distribution of marketed production of the major agricultural commodities. It does this through parastatal marketing boards for various commodities. At the beginning of each planning period, prices for various crops are announced, and market-oriented farmers make their planting decisions based on these prices. These prices should be set to promote the production of desired commodities, to reduce the risk of income fluctuations to farmers, to stabilize and reduce the final consumer prices of farm products, to promote equitable distribution of income, and to provide a rate of return sufficient to encourage individuals and firms to enter the agricultural industry.

The use of the price system to influence production depends on the response of farmers to changes in prices. The available evidence from our work shows that Zambian farm production responds to price (see chapter 4). In our model, we related value added in commercial and rural agriculture to the producer price of maize, the major agricultural commodity. We shall examine the effects of changes in that price (which is exogenously set in the model) on agricultural output, the rural-urban income distribution, and the shares of rural and commercial agriculture in total sectoral output.

A sustained 10 percent increase in the control-solution values of the producer price of maize is made for 1969 and 1990, and the model is solved with this

Table 7-11
Impact of a Sustained 10 Percent Increase in the Price of Maize

	Percentage Differences between Shocked and Control Solutions							
		Real Value Added			*Deflators*			
Years	*Real GDP*	*TAFF*	*CA*	*RA*	*GDP*	*CA*	*RA*	*Imports, SITC (0,1)*
1969	0.23	0.98	2.82	0.28	0.36	1.85	7.17	–0.34
1970	0.28	0.99	2.73	0.29	0.47	1.83	7.23	–0.17
1971	0.29	1.09	2.92	0.32	0.47	1.77	7.47	–0.11
1972	0.27	1.00	2.23	0.32	0.40	1.98	7.46	–0.08
1973	0.29	1.00	2.22	0.32	0.48	2.00	6.70	–0.10
1974	0.34	1.14	2.45	0.37	0.51	2.24	7.75	–0.58
1975	0.40	1.38	2.90	0.46	0.65	3.70	8.13	–0.76
1976	0.46	1.53	3.14	0.51	0.69	2.82	8.31	–0.98
1977	0.50	1.63	3.31	0.55	0.56	2.87	8.42	–1.10
1978	0.55	1.73	3.48	0.59	0.59	2.98	8.52	–1.41
1979	0.53	1.84	3.65	0.63	0.50	3.14	8.61	–0.96
1980	0.57	1.96	3.81	0.68	0.53	3.18	8.70	–1.06
1981	0.62	2.08	3.97	0.73	0.55	3.36	8.79	–1.30
1982	0.61	2.20	4.13	0.78	0.49	3.51	8.86	–1.03
1983	0.64	2.33	4.30	0.84	0.52	3.56	8.94	–0.96
1984	0.66	2.46	4.45	0.90	0.51	3.71	9.01	–0.96
1985	0.71	2.60	4.61	0.96	0.53	3.83	9.08	–1.07
1986	0.73	2.74	4.77	1.03	0.52	4.00	9.14	–1.05
1987	0.76	2.89	4.93	1.11	0.51	4.12	9.20	–1.06
1988	0.81	3.04	5.08	1.19	0.55	4.29	9.25	–1.11
1989	0.84	3.20	5.24	1.27	0.58	4.62	9.30	–1.22
1990	0.88	3.36	5.40	1.36	0.65	5.38	9.35	–1.17

Key: GDP = gross domestic product; TAFF = total agriculture, foresting, and fishing; CA = commercial agriculture; RA = rural (subsistence) agriculture; SITC (0,1) = food, beverages, and tobacco.

change. Table 7–11 presents the percentage differences between the control and the high maize price policy solution for selected variables. Figures 7–3 and 7–4 show graphical comparisons of the two solutions. Figure 7–3 shows the proportions of real GDP from agriculture in the two solutions. The proportion of agriculture in GDP is higher for the high-price solution, the difference between the two solutions increasing with time. The pattern of the share is the same in the two solutions, reflecting the similar pattern of maize prices. In figure 7–4, the proportion of real GDP from agriculture in the high-price solution is presented again in the top line; the two bottom lines represent agricultural incomes as a proportion of nominal GDP. The share of agricultural income is higher for the high-price solution, the difference in the two solutions is increasing with time, and the share of agricultural income is nominal GDP in the higher-price solution approaches the proportion of real output from agriculture.

Table 7–11 shows that increasing the price of maize increases real value added from agriculture, the commercial sector changing at a higher pace than the rural sector. This is consistent with the estimated relationship, showing that the commercial sector is more price-responsive than the rural sector. The price of

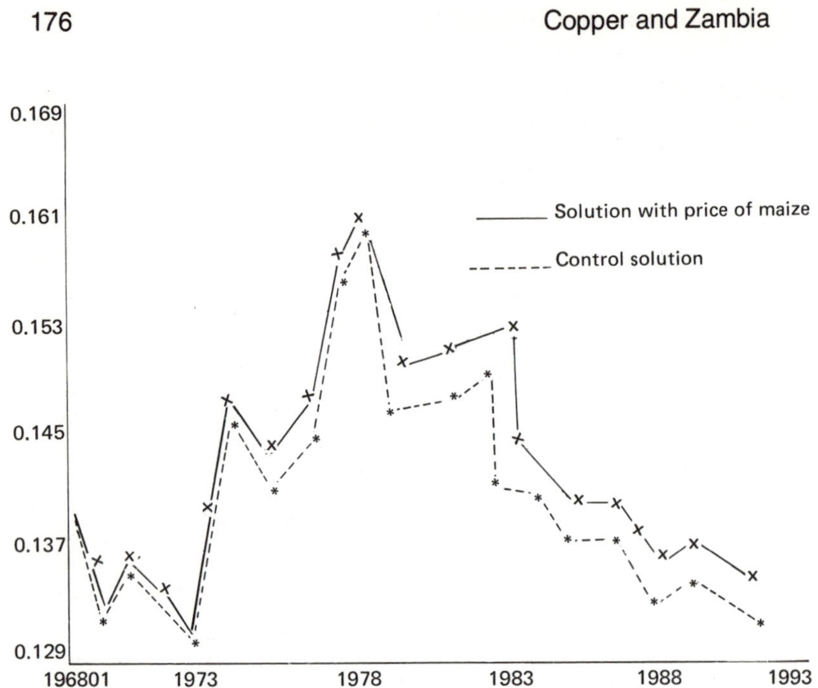

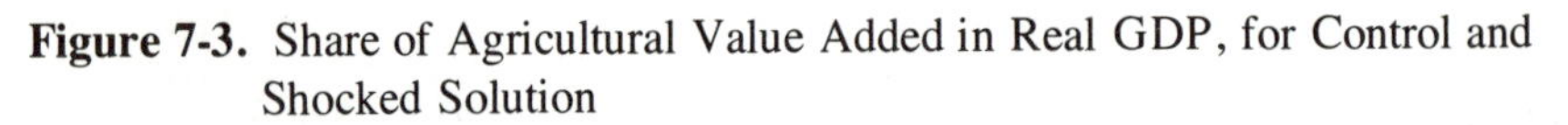
Figure 7-3. Share of Agricultural Value Added in Real GDP, for Control and Shocked Solution

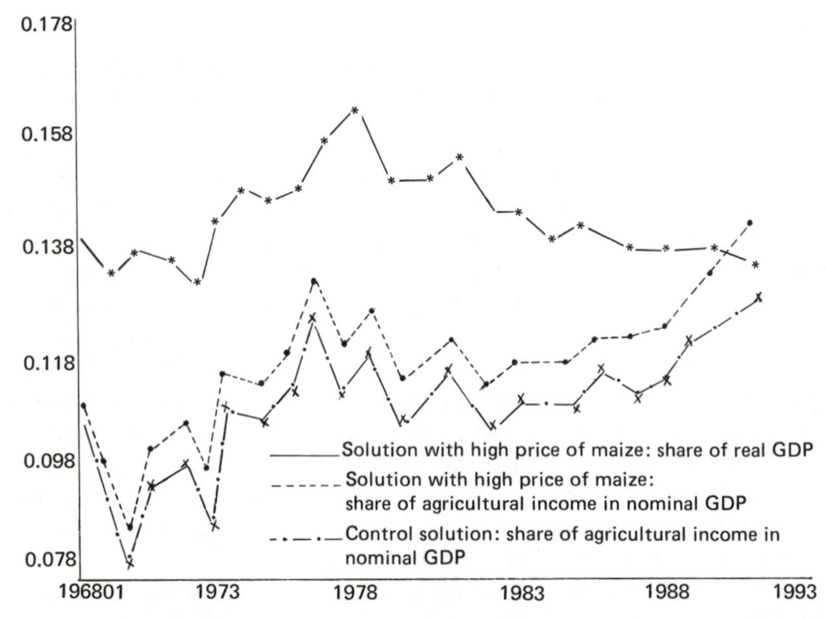

Figure 7-4. Share of Agricultural Value Added and Income in Real and Nominal GDP

output in both sectors rises, but the increase is higher for rural output than for commercial agriculture. The reason for the difference may be that the proportions of maize in the sectoral output are different. Rural agriculture accounts for about 80 percent of all maize produced,[14] but there is no available estimate of the proportion of maize in its total output. The distribution of agricultural output shifts toward more commercial production with the price increase, although the change is only slight. Real gross domestic product is higher in the high-price solution by less than 1 percent, and the deflator of GDP is also higher, by an even smaller percentage. The increased production induced by the price rise causes a reduction in the imports of food products.

The impact of the price increase on final consumer prices will depend on how much of the increased price is passed on to consumers by food distributors and processors. In the absence of any subsidies, almost all price increases will be passed on to consumers. Since the demand for food is likely to be inelastic, food distributors or retailers will pass on the price increases. The historical data indicate that the elasticity of the consumer price index with respect to the deflator of agricultural output is about 0.35 percent. Therefore, an increase in the price of agricultural production that raises the deflator by 10 percent will raise the CPI by 3.5 percent. In our simulations, the price deflator changes by a maximum of 8 percent in 1990; therefore, the maximum increase in the CPI will be about 3 percent.

The implications of these results are that with higher prices for maize, the incomes from agriculture are increased and the rural-urban terms of trade are improved for rural dwellers. This leads to a more equitable distribution of income. The increase in agricultural output is a positive step toward self-sufficiency in food commodities and increased exports of agricultural commodities. It also contributes to an increase in overall economic activity, although urban consumers will have to pay higher prices for their food. Insofar as redistribution of income toward the rural sector is a desirable goal, the inflation is tolerable. The impact of increasing the price of maize on real agricultural output might have been higher if the farmers were more responsive to price changes. The elasticities of commercial and rural agricultural outputs with respect to the price of maize were estimated to be 0.28 and 0.033, respectively. Rural agricultural output has a very low elasticity, because rural farmers market only a small proportion of their output.[15] Their primary interest is to produce sufficient food to feed their families; anything surplus to their needs may then be sold. Presumably, higher prices may induce traditional farmers to market a greater proportion of their output. At such prices, the change in relative prices will tend to induce farmers to substitute other goods for their own product in their consumption basket. If more rural farmers market their production because of this substitution incentive, the output of the sector may become more responsive to price. The income-distribution gains would be higher if farmers were more responsive to market trends; any farmer who does not sell any of his product will make no gains from higher prices.

Migration

The rural-to-urban migration has implications for agriculture. Traditional agriculture is labor-intensive and is a function of the number of people engaged in it. Migration from rural to urban areas drains the rural areas of young men, who are the most able prospective farmers. At the onset of urbanization, most migrants were young men. Recently, more young women are migrating, but the majority are still men.

Output in rural agriculture is responsive to the rural population. The estimated elasticity of real agricultural output with respect to the rural population is 0.33.[16] Thus, any policy that contributes to the slowing down or reversal of the rural-to-urban migration will stimulate rural agriculture. Differential returns to labor in rural and urban economic activities are regarded as an important factor affecting migration.[17] There are other contributing factors; for example, the educational system generally prepares students for urban-sector employment. Increasing returns in agriculture through higher prices for agricultural commodities may reduce the rate of migration and consequently increase agricultural output and reduce urban congestion.

The explanation of the migration process in our model makes the ratio of rural population to total population a function of the real urban wage rate, not the theorized relative returns from urban and rural employment. Therefore, our simulation of the changes in the price of maize does not capture any possible migration effects. We therefore perform a separate simulation, which explores the effect of reduced migration. In the control solution, the total population is assumed to grow at 2.5 percent, the urban population grows at about 5.5 percent, and the rural population shows no growth. In this simulation, the urban population is exogenized and is assumed to grow at a substantially lower rate, 3 percent. This gives a rural population growth rate of 2.2 percent. In addition, the real value added from the service sector and government spending on general services, which are partly dependent on the urban population, are exogenized.[18] The result of the simulation indicates a substantial increase in the output from rural agriculture. In 1977, the first year of change, real rural-agricultural value added is higher than the control-solution value by 0.6 percent, but, by 1990, the difference is 11 percent. The corresponding differences in the rural population are 1.7 percent and 35 percent. Total real GDP is higher for the low-migration solution, and real private consumption is higher.

The simulation assumes that the urban population is sufficient to meet the demand for labor in the urban sector. This condition is likely to be met, as employment in the urban sector grows at less than the assumed growth rate of the urban population. The attractive aspect about increasing production in rural agriculture through migration policies and agricultural pricing is that it does not compete for scarce resources, such as manpower, with the other sectors. If higher prices of agricultural commodities induce a reduction in the rural-to-urban migration, there will be a movement up the supply curve and a shift of the supply curve to the right, both effects resulting in increased supply.

There are ways of achieving higher production in agriculture other than the ones that have been discussed. Increased productivity can be obtained by the use of improved farm implements, higher-yielding varieties of crops, better organization of farm production, such as cooperatives, higher utilization of available farm capital equipment, increased use of fertilizers and irrigation, and extension services. These methods are especially important for the long-term performance of the industry. In addition, rural-to-urban migration can be reduced by policies other than increasing returns in agriculture, such as improvement of health and educational facilities, better supplies of electricity and water, and improved transportation services in the rural areas. We have concentrated on the role of traditional economic incentives, which induce those individuals and firms already in the industry to work more and those outside the industry to enter it.

Notes

1. See chapter 5. The results are shown in tables 5–6 and 5–7.
2. See M. Friedman, "Monetary Policy in a Developing Economy," in P. David and M. Reder (eds.), *Nations' and Households' Economic Growth* (New York: Academic Press, 1974).
3. See R. Jolly and M. Williams, "Macro-Budget Policy in an Open Export Economy: Lessons from Zambian Experience," *Eastern Africa Economic Review* 4 (December 1972).
4. This policy reduces uncertainty and costs of adjustment in the economy in which government spending is a large part of all expenditures (see chapter 2).
5. There is no significance in the fact that government expenditures are raised by 10 percent and copper prices are lowered by 10 percent.
6. See chapter 3 under Price Fluctuations and Inflation.
7. At the time of this writing, the government owned 51 percent of RCM and 60 percent of NCCM.
8. Although the government has effective control, minority partners could still block any proposal by simply not showing up at the board meeting, thus denying the board a quorum.
9. See M. Faber and G. Potter, *Towards Economic Independence: Papers on the Nationalization of the Copper Industry in Zambia* (Cambridge, England: Cambridge University Press, 1971), page 114.
10. See *Appendices to Explanatory Statement of Roan Selection Trust Ltd.*, 30 June 1970, App H, p. H–1.
11. See chapter 3. In the profit-maximizing case, copper output is given by equation 3.10; in the output-maximization case, equations 3.11 to 3.15 determine output.

12. See L.R. Klein and R. Summers, *The Wharton Index of Capacity Utilization,* Economics Research Unit, University of Pennsylvania, 1966.

13. The assumption is that cumulative output is related exponentially to grade, so the inverse is estimated.

14. See Republic of Zambia, *Census of Agriculture, 1970–71 (Second Report).*

15. The *Census of Agriculture, 1970–71 (Second Report)* shows that, for the 1970–71 agricultural year, noncommercial farmers sold or bartered 36 percent of their output of maize.

16. See chapter 4.

17. See J. Harris and M. Todaro, "Migration, Unemployment and Development: A Two Sector Analysis," *American Economic Review* 60 (March 1970): 126–142.

18. Although these formulations may be adequate for small endogenous changes in the urban population, they will not be appropriate for large structural changes in population dynamics.

Conclusions

The objective of this chapter is the integration of the findings from all the earlier analyses. The implications of these findings for economic policy will be analyzed. The analysis will summarize the findings on the impact of the copper industry on the economy and the efficacy of fiscal and monetary policies and commercial policies in the copper sector in counteracting the negative impact of copper-price fluctuations and promoting economic growth and welfare.

The Impact of the Copper Industry

General Macroeconomic Effects

The multiplier tests show that changes in the price of copper produce changes in the real output in the copper sector, but the induced change in the real output and employment in other sectors is small in the short term. The major short-term effect of copper-price changes is on the aggregate price level, which rises when the price of copper rises because of induced changes in the money supply. The deflators of output in the copper and noncopper sectors and the consumer price index all rise with an increase in copper prices. This response to copper price changes is a result of the very low price elasticity of supply of the noncopper sectors and the weak backward linkages between the copper and noncopper sectors. Changes in the price of copper have an immediate impact on copper-industry profits, the balance of payments and international reserves, and government revenues. Exogenous changes in the level of copper production affect copper-sector employment, balance of payments, and reserves; the short-term effect on real product in noncopper sectors is even weaker than the price effect. Changes in copper production do affect the aggregate price level.

The short-term effects of copper-price or production changes do have some long-run effects. An increase in the price of copper raises copper-industry profits. Compensation for labor in the copper industry responds to this increase in profits, usually with a lag. Since wages and salaries in noncopper sectors tend to move with copper wages, the effect of the increase in the price of copper is to raise wages and salaries in the economy. In this way, changes in copper prices have a long-run effect on the aggregate price level in the economy. Short-term changes in copper production or prices affect the level of international reserves, which affects real imports in subsequent periods by changing the ability to import. In the long run, changes in the secular trend of copper prices produce significant induced effects on other sectors' real product and employment by changing real investment and government spending.

The major channels through which the copper industry affects the economy in the long term are government revenues and the industry's impact on both private and public investment. Government investment in the private sector in the form of loans and equity holdings depends on the amount of government budget deficit or surplus and has a significant effect on private noncopper-sector investment. The budget deficit or surplus is affected by copper prices through government revenues from the industry.

The level and variability of foreign-exchange earnings or reserves did not have any significant macroeconomic effects in the model. We tested for the effect of the level and variability of foreign reserves on private-sector (copper and noncopper separately) real capital formation, but none was significant. The level of foreign reserves affected only some categories of imports of goods. We also tested for the effect of imports of goods on investments and real sectoral outputs and found that the effects were insignificant, except for small effects of imports of crude materials on value added from the manufacturing sector. There have been reports of shortages of imported inputs affecting production and investment in mining and other sectors;[1] these shortages have been sporadic and have been associated with transportation delays and inefficiencies in the functioning of the import-licensing system rather than with the availability of foreign exchange.

The major reason for the insignificance of the effect of foreign exchange on real output and investment lies with the economic realities of the period within which the model was estimated. Most of that period corresponded to a boom in the copper market, resulting in foreign-exchange earnings by Zambia that were, at most times, in excess of its needs. During this period, the major constraint on economic activity was manpower; the availability of foreign exchange made it possible to import significant numbers of skilled persons from abroad. Government policy also gave favorable treatment to imports of intermediate and investment goods; any reductions in imports were generally borne by consumption goods.

Changes in copper prices and the level of activity in the copper sector had an insignificant impact on the agricultural sector in the short run. Since prices of the major agricultural commodities are set by government, the price effect of copper price-output changes is not transmitted to agricultural prices except to the extent that the induced inflation or deflation makes the government adjust agricultural prices in the interest of equity. In the long run, copper-sector activities affect agriculture through investment in commercial agriculture and migration from rural to urban areas, induced by changes in formal sector wage rates, in which the copper sector is the trend setter. Migration from rural to urban areas has a negative effect on rural agricultural production.

Effects of Export Instability

It was shown that changes in copper prices have historically been the major cause of the instability of export revenues, accounting for over 75 percent of the deviation of revenues from trend growth path. It was found that export instability engendered by copper-price fluctuations did not have any substantive impact on aggregate output and employment in the economy. However, fluctuation of copper prices is the source of real and nominal instability in the economy, as measured by the trend-adjusted variance of real gross domestic output, domestic price level, and other major variables. In addition, fluctuation of copper prices results in higher domestic price inflation.[2]

Export instability has no significant effect on real economic activity in Zambia, for several reasons. The ownership structure of the industry protected the economy from price fluctuations, as the foreign shareholders and the government absorbed the impact of the fluctuations. For most of the period within which the model was estimated, revenues did not impose a binding constraint on government spending; although expenditures responded to revenues, the response was small. When revenues did not meet planned expenditures, the government had substantial reserves and borrowing capacity at home and abroad to draw on. While government investment in the private noncopper sector responded strongly to budget surplus or deficit, the private sector had access to commercial bank credit at low interest rates to smooth the fluctuations in the receipts from the government and to keep their investments at planned levels. Hence, both the private sector and the government were able to maintain their spending in the face of fluctuations of revenues.

The response of the Zambian copper industry to copper prices was weak in the short run, and so real activity in the copper sector did not suffer much from price fluctuations. The low price response of the Zambian industry can be attributed to several factors. Zambian mining was for the most part a growing industry, with high-grade copper ores and low-cost operations, and so capacity constraint was a significant factor in the level of production. The lack of speedy access to world markets for required imports of investment and intermediate input goods, the lack of skilled labor, and the cautious investment behavior of the foreign-owned mining companies tended to slow changes in production capacity.

These factors shielded Zambia from copper-price fluctuations. In the absence of these extenuating circumstances, however, the story would be very different. It may well be that with the new ownership arrangements in the copper industry and the tighter financial resource situation, due in part to low copper prices and in part to Zambia's growing needs, managing the economy under fluctuating export and government revenues will be more difficult and more costly. When financial resources were plentiful, deviations of actual performance from planned performance involved few real costs, but as resources become scarce, such deviations are likely to be costlier. With a reserve of borrowing capacity, foreign-exchange commitments in excess of planned use may be met by borrowing in capital markets, without affecting interest costs. When such borrowing capacity is not available, any borrowing can only be accomplished, if at all, at much higher costs. The same would apply to domestic borrowing; an unexpected deviation in government borrowing needs will most likely be met by monetary expansion, and unanticipated changes in money supply lead to unanticipated inflation.

Income-Distribution Effects

Copper mining, as the first large-scale industry in Zambia, set the trend for earnings in formal employment. Initially, the industry relied on imported labor and paid salaries that were competitive with those in the United Kingdom and South Africa, its major sources of supply. In time, more Africans were employed, but at wages much lower than those of the expatriates, resulting in the demand by the African unions and preindependence nationalists to narrow or even close the gap between the expatriates and locals. Although parity has never been achieved, the effort has nonetheless yielded the mining employees wages that are much higher than they could obtain in alternative employment.

Wages in other modern formal sectors are lower than those in mining, but our analysis shows that these wages tend to follow the trend of mining wages. The mining companies also set the trend for the compensation of skilled and professional Zambian employees (based on compensation of expatriates), which the government and other industries were compelled to match in order to attract this scarce resource. The copper industry has therefore been responsible for raising and maintaining Zambia's urban-sector wages at high levels, compared to alternative incomes in the agricultural and informal sectors.

The impact of high wages is that Zambia has a very high rate of urbanization; rural dwellers are attracted by the high incomes and better services in the urban areas. The high wages also affect the amount of services and investment funds the government can provide as it increases the cost of labor-intensive government services. The formal-sector employees can substitute other goods and services with their additional income, which the rural and informal-sector workers cannot do. High wage payments may be one of the factors responsible for the tendency toward greater capital intensity in the formal sector, thus reducing employment opportunities and concentrating income with the owners of capital.

The inflationary bias induced by copper-price fluctuations may have an adverse effect on income distribution. Unanticipated inflation tends to make the problem of achieving a desired pattern of income distribution through the use of policy more difficult, as the inflation could negate the intent of the policy. In the Zambian case, the effect of the inflation is to distribute income to the government and the urban sector. The informal and rural sectors possess neither the organization nor the market or political power to protect themselves from inflation.

Policy Issues

Fiscal and Monetary Policies

Model simulations show that fiscal and monetary policies can be used to partially offset the effects of fluctuations of copper prices. When the price of copper falls, an increase in government spending can compensate for the loss of output and employment resulting from the decline in copper prices. The inflationary effect of an increase in copper prices can be reduced by sterilizing the induced inflows of foreign exchange or by reducing government spending below the level that would be dictated by the higher level of revenues if the government reacted passively. The most potent policy for reducing the induced inflationary effect is the control of the money supply.

The use of fiscal policy to offset the effect of a decline in copper prices is dependent on the availability of such resources as manpower and domestic and foreign financial reserves. A decline in copper prices usually leads to a decline in real economic activity in the copper sector, which is unlikely to release sufficient skilled labor and managerial personnel to put new projects in place in other sectors. To increase government spending when copper prices are falling implies the accumulation of fiscal spending and balance-of-payments deficits. Unless reserves were accumulated during past periods of higher copper prices or the government has additional borrowing capacity, an offset policy will be difficult to accomplish. In the long run, a policy of increasing expenditures when copper prices are falling would have to be matched with one of decreasing expenditures when prices are rising, if expenditures and revenues are to be in long-run balance. Therefore, a policy that sets government spending on long-run considerations may be more appropriate.

Simulation results using the econometric model of Zambia indicate that stabilizing government expenditures on a long-run growth path stimulates gross output, reduces the price level and its variance, but increases the instability of real variables. A stable expenditure policy will make the coordination of economic planning easier, once the long-term growth paths of expenditures have been determined and made known. It will also have the beneficial effect of reducing uncertainty in the economy, in which the government plays a very active role. The major obstacle to implementing a policy of stable spending is the determination of future revenues on which the growth rates of spending are to be based. The price of copper is the most difficult variable to forecast in the determination of future revenues. In this case, the government, rather than investing substantial resources in an effort to forecast copper prices, should make use of the forecasts of multilaterial bodies such as the World Bank and the United Nations agencies.

A long-term spending plan should have some built-in flexibility so that it can be adjusted to changing long-term market conditions or unusual short-term conditions. Such flexibility could be achieved by ordering future projects in terms of priority and dropping low-priority budget items if market conditions or expectations change.[3] In addition, flexibility could also be achieved by exploring sources of financing of government or balance-of-payments deficits in advance.

Monetary policy was shown to have little impact on real economic activity in the short run but to affect the price level. It was also found that stabilizing the money supply would reduce the level and variance of the consumer price index but would not reduce the real instability induced by copper-price fluctuations. Stabilizing both the money supply and government spending reduces the level and variance of the money supply but increases real instability. The increase in real instability from domestic price stabilization stems from the decrease in the covariance of the domestic price level and the nominal shocks from copper-price changes when the domestic price level is stabilized. This stabilization increases the variance of the real copper-price shocks. The result implies that the attainment of both real and price stability with fiscal and monetary policies in the face of fluctuations of copper prices is an extremely difficult task.

In the long run, the money supply affects growth of the economy through its effect on investment. Thus, the objective of monetary policy should be to provide the economy with sufficient liquidity to finance investments while maintaining price stability. A stable growth rate in money supply would achieve this objective.

An important aspect of monetary policy, which deserves attention, is the setting of interest rates. Historically, Zambian interest rates have been low and have reflected the changes in the price level. Commercial-bank deposit rates on savings accounts varied from 3 percent in 1966 to 5.5 percent in 1976.[4] The corresponding inflation rates were 10 percent and 20 percent. Similarly, average treasury bill rates were 3.8 percent in 1966 and 4.3 percent in 1976, and commercial-bank minimum-lending rates were 6.5 percent and 8 percent in 1966 and 1976, respectively. The result of low interest rates is that capital is rationed; the likely recipients of bank credit are customers who are favored for personal, political, or business relationships. Hence, the role of interest rates in improving the allocation of investments is lost. Low interest rates depress savings and encourage consumption and the holding of idle money balances. Easy and inexpensive credit to favored customers encourages them to move toward capital-intensive techniques, to the detriment of employment.

Copper-Industry Policies

Model simulations have shown that a policy of varying copper production with the objective of maximizing short-run profits is dominated by a policy that uses available capacity to achieve maximum output. In these simulations, maximum-output solutions gave higher copper-industry output, profits, and employment of both Zambians and non-Zambians. Higher copper production and exports lead to higher foreign revenues and reserves. The increased activity in the copper sector induces increased investment, employment, and output in the other sectors in the economy. These effects of output maximization continued to obtain on a smaller scale when changes in grade of copper ore are made to respond to changes in cumulative output. However, because of the aggregative nature of the analysis, the effect of output maximization on industry profits and its induced effect on the economy is not very robust. The major merit of output maximization is the direct effect on industry employment and improved foreign-exchange earnings.

The policy of output maximization in the Zambian copper industry reduces the real instability of the value added from the copper sector, which leads to the reduction of the trend-adjusted variance of the gross domestic output. However, it increases the level and the variance of the domestic price level, although this can be reduced and stabilized by stabilizing the money supply.

The kind of policy the government and the industry adopt will generally depend on long-term government-development strategy for the industry. If the intention is to develop the industry in partnership with multinational companies, the profit-maximization policy ought to be the ruling policy. This will attract more foreign risk capital and management and will probably speed up the development of the industry and achieve the desired employment and foreign-exchange goals. Alternatively, if the government is to bear all future financial risk in the development of the industry, a policy of output maximization would be best for the economy.

The copper industry is an important source of revenue for the government; the present taxation system raises this revenue through a tax on profits. If an output-maximizing policy were to reduce profits, government revenues from mining will fall. However, the induced effects of output maximization will increase revenues from other sources, for example, increased taxes from wage income, due to higher employment in the copper and other sectors. This may not compensate for the loss of direct mineral revenues, since the marginal tax rate on other forms of taxes is less than the marginal tax rate on mineral profits. Lack of profitability in the copper industry may have an adverse effect on the morale and productivity of the employees in the industry, because only a profitable enterprise can provide and fulfill the expectation of higher wages.

The best policy for the industry may be a flexible combination of output and profit goals. The implementation of this combined goal may be reflected in the incentive-based compensation of management. The Zambian copper industry has a complete separation of ownership and control. The owners are the Zambian people and some foreign interests, and the managers are individual Zambians and expatriates. There is no reason for management to maximize any particular objective unless there is some incentive to do so, especially since a competitive market for managers does not exist. A performance-based incentive scheme for the managers could be a compensation scheme based on a combination of sales of copper and profits, with a provision that the weights could be adjusted after consultation between the owners and management. Such an incentive scheme could also be extended to all employees, which would not be new to Zambian mining industry, as it used to pay an annual bonus to its employees based on profitability.

Suggestions for Further Work

The effort of building the model of Zambia was hampered by the lack of data for many important economic variables and the shortness of the available series. As more data become available, both in scope and length of time series, further disaggregation and reestimation of the model would probably improve the behavior of the model. For example, more data on sectoral investment and capital stocks would be an important addition to the sectoral disaggregation of the model and would improve the determination of sectoral value added. It would also make it possible to determine to a greater extent how changes in copper prices affect the sectoral distribution of investment and output. Data on foreign investment in the Zambian economy may be an important determinant of gross investment, but they are not readily available. If they were, they would provide clues on how government economic reforms have affected foreign investment and thus economic growth.

The Zambian economy has undergone several structural changes, from the effects of political independence to the various economic reforms the government has put into place. It is likely that such changes have had an effect on the behavior of agents in the economy; such changes ought to be reflected in the coefficients of the model. We have attempted to capture such changes with dummies, but we have been restricted by the shortness of the data series. With longer data series, it may be possible to capture more of the impact of these structural changes. In the projection of the economy from 1977 to 1990, we have basically assumed that the structure will remain as it was in 1976. The model can be used to study the impact of major structural changes, but we have not done so.

The scope of the model has been primarily on the urban-modern sector. With more data and other resources, the model could be extended to cover the economic activities in the rural and informal economies more adequately. This would make it possible to determine the impact of the copper sector on the rural economy. The channel of this effect would be mainly through government expenditures on rural development in addition to the impact on rural-urban migration.

Notes

1. For example, see Republic of Zambia, *Economic Report 1973,* Ministry of Planning and Finance, Lusaka, January 1974, pp. 192, 251.

2. The effect of copper-price fluctuations on the instability of domestic variables and on the level and variance of the domestic price level is analyzed in C. Obidegwu, "The Impact of Copper Price Movements and Variability on the Zambian Economy: A Dynamic Analysis," Ph.D. dissertation, University of Pennsylvania, 1980.

3. One such proposal of priority budgeting for Zambia is suggested in R. Jolly and M. Wiliams, "Macro-Budget Policy in an Open-Export Economy: Lessons from Zambian Experience," *East African Economic Review* 4 (December 1972).

4. See Republic of Zambia, *Monthly Digest of Statistics,* various issues.

Appendix A Variable Definitions and Model Identities

The major sources of data were:

1. The World Bank, *World Tables* 1976.
2. Republic of Zambia, *Monthly Digest of Statistics,* Central Statistical Office, Lusaka, several issues.
3. The International Monetary Fund, *International Financial Statistics,* several issues.
4. The *Zambian Mining Year Book,* several issues.
5. *Annual Reports* of the Zambian mining companies, several issues.

For the variables, unless otherwise specified:

Employment is in thousands of man years.

Earnings are in kwacha, except total earnings.

Copper-production figures are in thousands of metric tons (tonnes).

Unit costs in copper mining are in kwacha per metric ton.

All other variables are in millions of kwacha, except variables defining price indexes or deflators, ratios, and shares. Real values are in 1965 kwacha.

The variable listing contains mainly the exogenous and behavioral variables. The identities can be obtained by using the definitions in the list of model identities or in the text. The list of identities contains those that are not available in the text.

Definitions of Endogenous Variables

AVWAGE	Average compensation in the copper industry: all employees
BCGC	Government final-consumption expenditure
BCGC65	Government final-consumption expenditure in constant 1965 prices
BCPC	Personal final-consumption expenditure
BCPC65	Final personal-consumption expenditure at constant 1965 prices
BETA	Employees' share of output in the copper industry
BPC	Current-account balance
BPCA	Capital-account balance (net)

BPEXG	Exports of goods (FOB): balance of payments
BPNECF	Change in net foreign assets (+ = increases)
BPT	Balance of trade
CFC	Consumption of fixed capital (depreciation)
CGBS	Central-government recurrent budget surplus
CGCD	Central-government recurrent revenue from customs duties
CGCE	Total central-government capital expenditure
CGDC	Central-government direct capital expenditure
CGES	Central-government recurrent expenditure on economic services
CGEST	Central-government recurrent revenue from excise and sales taxes
CGGS	Central-government recurrent expenditure on general services
CGIT	Central-government recurrent revenue from income taxes
CGLIC	Central-government capital-expenditure loans and investments
CGLIC65	Central-government capital expenditure: loans and investments in 1965 prices
CGMR	Central-government recurrent revenue from mining
CGND	Central-government overall budget deficit (−)
CGRD	Central-government other recurrent revenues
CGRE	Central-government net recurrent expenditure
CGRR	Central-government recurrent revenue
CGSS	Central-government recurrent expenditure on social services
CUQCZM	Copper production in copper content of concentrates
CUQCZMP	Production capacity in copper mining
CUR	Capacity-utilization rate in the copper industry
CVEXCOP	Coefficient of variation of copper exports from moving average
DCGRR	Change in government current revenues
DFINCI	Change in real final demand
DMSQM	Change in supply of money and quasi-money
DNNGD	Change in net claims on the government
DNPSD	Change in net claims on the private sector
EXCH	Exchange rate: pounds sterling/kwacha
EXCOP	Exports of copper
EXCOPQT	Exports of refined and unrefined copper (tonnes)
EXCOP65	Merchandise exports at constant 1965 prices: copper
EXCPRICE	Realized export price of copper
EXGD	Total exports of domestic merchandise (FOB)

EXGS	Total exports of goods and services
EXGS65	Total exports of goods and services, constant 1965 prices
FINCI	Final demand, real
FINCIN	Final demand
IGCA65	Real gross fixed capital formation in commercial agriculture
IGCON65	Real gross fixed capital formation in construction
IGFC	Gross fixed capital formation
IGFCP	Private gross fixed investment
IGFCPNM65	Private noncopper gross fixed investment in 1965 kwacha
IGFCP65	Private fixed investment in 1965 kwacha
IGFC65	Gross fixed capital formation, constant 1965 prices
IGMAN65	Real gross fixed capital formation in manufacturing
IGMQ65	Real gross fixed capital formation in mining and quarrying
IGSER65	Real gross fixed capital formation in services
IGTC65	Real gross fixed capital formation in transportation and communications
INCHS	Increase in stocks
INCHS65	Increase in stocks, constant 1965 prices
INDTAX	Indirect taxes
KCMIN	Capital formation in copper industry
KCMINREAL	Real investment in copper mining in constant 1970 import prices
KCMIN65	Real investment in copper mining
KSCA65	Real capital stocks in commercial agriculture, end of year
KSC65	Real capital stocks in construction
KSMAN65	Real capital stocks in manufacturing, end of year
KSMQ65	Real capital stocks in mining and quarrying, end of year
KSS65	Real capital stocks in the services sector
KSTRCOM65	Real capital stocks in transportation and communications
KS65	Real capital stocks, total all sectors, end of year
LEAGZ	Employment in agriculture, forestry, and fishing: Zambians
LECZ	Employment in construction: Zambians
LEEXPR	Expatriate employment in noncopper sectors
LEIND	Total employment, all industries: Zambians and non-Zambians
LEINDN	Total employment in all industries: non-Zambians
LEINDZ	Total employment in all industries: Zambians

LEMANZ	Employment in manufacturing: Zambians
LEMAQN	Expatriate employment in mining and quarrying
LEMAQZ	Employment in mining and quarrying: Zambians
LESERZ	Employment in the services sector: Zambians
LETACZ	Employment in transportation and communications: Zambians
LKKLR	Log of capital-labor ratio in copper mining
LOLR	Log of output-labor ratio in copper mining
LVLOR	Log of labor-output ratio in copper mining
MCAPRES	Proxy variable for ability to import (NINTR(-1)/PMGS)
MCH	Merchandise imports by SITC categories: chemicals
MCH69	Merchandise imports by SITC categories in 1969 prices: chemicals
MGFOB	Total merchandise imports (FOB)
MGS	Total imports of goods and services
MGS65	Total imports of goods and services, constant 1965 prices
MGTT	Imports of merchandise, less chemicals and machinery and equipment
MG65	Merchandise imports, in 1965 kwacha
MG69	Merchandise imports, in 1969 kwacha
MMAG	Merchandise imports by SITC categories manufactured goods
MMAG69	Merchandise imports by SITC categories, in 1969 prices: manufactured goods
MMATE	Merchandise imports by SITC categories: machinery and transportation equipment
MMATE69	Merchandise imports by SITC categories, in 1969 prices: machinery and transportation equipment
MMMA	Merchandise imports by SITC categories: miscellaneous manufactured articles
MMMA69	Merchandise imports by SITC categories, in 1969 prices: miscellaneous manufactures
MSQM	Money supply plus quasi-money
MSS	Imports of services
MSS65	Imports of services, in 1965 kwacha
MXCOP	Moving average of copper exports
M01G	Merchandise imports by SITC categories: food, beverages, and tobacco
M01G69	Merchandise imports of food, beverages, and tocacco in 1965 kwacha
M24G	Imports of crude materials excluding fuel: SITC categories 2 and 4

M24G69	Imports of crude material excluding fuels, in 1965 kwacha
M25G	Imports of merchandise in SITC categories 2 to 5
M25G69	Imports of merchandise in SITC categories 2 to 5, in 1969 kwacha
M3G	Merchandise imports by SITC categories: electricity and mineral fuels
M3G69	Merchandise imports by SITC categories, in 1969 prices: electricity and mineral fuels
NFORA	Net foreign assets
NINTR	International reserves, end of period
NNGD	Net claims on the government
NPSD	Net claims on the private sector
PBCGC	Deflator for government final-consumption expenditure
PBCPC	Deflator for personal-consumption expenditure
PC	Consumer price index, 1975 = 100.0
PGDP	Deflator for gross domestic product
PGDPNM	Deflator for gross domestic output in nonmining sectors
PIGFC	Deflator for gross fixed capital formation
PMG	Unit value of imports of goods (FOB)
PMGS	Deflator for imports of goods and services
PMMAG	Deflator for imports of manufactured goods
PMMATE	Deflator for imports of machinery and transportation equipment
PMMMA	Deflator for imports of miscellaneous manufactures
PMSS	Deflator for imports of services
PM01G	Deflator for imports of food, beverages, and tobacco
PM24G	Deflator for imports of crude materials excluding fuels
PM3G	Deflator for imports of fuels
PM5G	Deflator for imports of chemicals
POPR	Estimated rural population
POPU	Estimated total urban population (large and small urban areas)
PROF	Urban-sector operating surplus, less copper-mining profits
PXAFF	Deflator for output in agriculture
PXAFFCS	Deflator for output in commercial agriculture
PXAFFS	Price deflator for output in rural agriculture
PXCONST	Deflator for output in construction
PXGC	Unit value of exports of copper (1965 = 1.00)
PXGS	Deflator for exports of goods and services
PXMANF	Deflator for output in manufacturing
PXMINQ	Deflator for output in mining and quarrying
PXSER	Deflator for output in the service industry

PXTRCOM	Deflator for output in transportation and communications
SDISC	Statistical discrepancy
SDISC65	Statistical discrepancy, constant 1965 prices
SHMQ	Sectoral share of real gross fixed capital formation: mining and quarrying
SHSR	Sectoral share in nonmining real investment: services
TLC	Total labor costs in copper mining
TOT	Terms of trade
UBCP	Real urban per capita consumption
UBC65	Total urban consumption, in 1965 kwacha
ULCNM	Unit labor costs in nonmining sectors
VGOREAL	Total capital stock in copper mining
VGP	Copper-industry pretax profits
VGPNET	Copper-industry net after-tax profits
VGPNET65	Calculated real after-tax profits in copper industry
VGPTAXB	Calculated tax base of copper-mining companies
VIMNC	Total cost of intermediate inputs in copper mining
VKZPRI	Realized price of copper sales, in kwacha per tonne
VLEO	Local employees in open-pit mining
VLORO	Labor-output ratio in open-pit mining
VLORU	Labor-output ratio in underground mining
VMTAX	Mineral tax payable in mining companies
VPCE	Consumer price index: European
VQCZM	Finished production of copper from new ore
VSXCOP	Sales revenue from copper mining
VTAXEX	Export taxes payable by mining companies
VTAXROY	Royalties payable by mining companies
VTEE	Total expatriate employees in copper mining, in manyears
VTESP	Copper sales
VTLE	Total local employees in copper mining, in manyears
VTOPC	Total operating costs in copper mining, including transportation
VTOTP	Total copper production (finished output)
VTTCOST	Total transportation costs in copper mining
VUMC	Real per-unit input of intermediate goods in copper mining
VUMCO	Real per-unit material input to open-pit mining operation
VUMCU	Real per-unit material input to underground mining operation
VXCOP	Value added to economy by copper sector
VXNCCM	Copper production by NCCM (from 1970; otherwise, VXTO)

VXRCM	Copper production by RCM (from 1970; otherwise, VXTU)
VXTO	Finished production from open-pit mining
VXTU	Finished production from underground mining
VZPRICE	London Metal Exchange price of copper, in kwacha per metric ton
WAGEIND	Index of wages in copper mining (1965 = 100.0)
WGAAFFZ	Average annual earnings in agriculture, forestry, and fishing: Zambians
WGACZ	Average annual earnings in construction: Zambians
WGAEXPR	Wage rate for expatriates in noncopper sectors
WGAMQN	Average annual earnings in mining and quarrying: non-Zambians
WGAMQZ	Average annual earnings in mining and quarrying: Zambians
WGAMZ	Average annual earnings in manufacturing: Zambians
WGASZ	Average annual earnings in the services sector: Zambians
WGATCZ	Average annual earnings in transportation and communications: Zambians
XAFF	Output in agriculture, forestry, and fishing
XAFFCS	Output in agriculture, forestry, and fishing: commercial sector
XAFFCS65	Real output in agriculture, forestry, and fishing: commercial sector
XAFFS	Output in rural agriculture
XAFFS65	Real output in rural-subsistence agriculture
XAFF65	Real output in agriculture, forestry, and fishing
XCONST	Output in construction
XCONST65	Real output in construction
XDOTA	Ratio of copper exports and its moving average
XGDCOP	Contribution of copper in net domestic product
XGDMC	Gross domestic product at market prices
XGDMC65	Gross domestic product in constant 1965 prices
XGNP65	Gross national product in constant 1965 prices
XMANF	Output in manufacturing
XMANF65	Real ouput in manufacturing
XMINQ	Output in mining and quarrying
XMINQ65	Real output in mining and quarrying
XSER	Output in the services sector
XSER65	Real output in the services sector
XTRCOM	Output in transportation and communications
XTRCOM65	Real output in transportation and communications
YIEX	Investment income paid abroad

YIM	Investment income received from abroad
YN	National income at market prices
YNIEX	Net (investment) income paid abroad, NA
YNODEX	Net other current (government and private) transfers from abroad
YPROF	Total operating surplus
YPROFD	Urban operating surplus, less investment income paid abroad
YPROFR	Urban operating surplus, less mining profits and indirect taxes
YPROFRPC	Real net urban operating surplus per capita
YW	Compensation of employees
YWC	Cash earnings of employees
YWD	Disposable wage income, less income transfers abroad
YWDR	Real net urban disposable income
YWDRPC	Real net urban disposable income per capita
YWEXPR	Total wage earnings of expatriate employees
YWO	Other noncash earnings (employers' contributions to pension funds, etc.)
ZMCOST	Weighted average cost of mining operations

Definitions of Exogenous Variables

BPEO	Errors and omissions: balance of payments
BPPCG	Net public-capital inflows
BPPCO	Other capital inflows (net)
BPPCX	Net private-capital inflows
CGMRADJ	Adjustments to government mineral revenues
CGREO	Other government expenditures
CGRPMT	Repayments of loans from central government
DMUFULIRA70/1	Dummy variable for Mufulira mine accident (1970–1971 = 1.0)
DSZM6202	Dummy variable for mining-industry labor strike (1962 = 1.0)
DZEXTAX	Dummy variable for export taxes (1966–1969 = 1.0)
DZF	Dummy variable, DZF = 1 up to and including 1965
DZHALF73/74	Dummy variable (1973 = 0.5, 1974 = 1.0)
DZHALF75/76	Dummy variable (1975 = 0.5, 1976 = 1.0)
DZI71/72	Dummy variable (1971–1972 = 1.0, otherwise 0)
DZ51/69	Dummy variable (1951–1969 = 1.0, otherwise 0)
DZ58	Dummy variable (1958 = 1.0)
DZ64/66	Dummy variable (1964–1966 = 1.0)
DZ66	Dummy variable (1966 = 1.0, otherwise 0)

DZ66/68	Dummy variable (1966–1968 = 1.0)
DZ66/69	Dummy variable (1966–1969 = 1.0)
DZ67	Dummy variable (1967 = 1.0, otherwise 0)
DZ67/68	Dummy variable (1967–1968 = 1.0)
DZ68	Dummy variable (1968 = 1.0, otherwise 0)
DZ69	Dummy variable (1969 = 1.0)
DZ69/70	Dummy variable (1969–1970 = 1.0, otherwise 0)
DZ70/71	Dummy variable (1970–1971 = 1.0)
DZ70/76	Dummy variable (1970–1976 = 1.0)
DZ71	Dummy variable (1971 = 1.0)
DZ71/72	Dummy variable (1971–1972 = 1.0)
DZ72	Dummy variable (1972 = 1.0, otherwise 0)
DZ72/76	Dummy variable (1972–1976 = 1.0)
DZ73	Dummy variable (1973 = 1.0)
DZ73/74	Dummy variable (1973–1974 = 1.0)
DZ74	Dummy variable (1974 = 1.0)
DZ75	Dummy variable (1975 = 1.0)
DZ76	Dummy variable (1976 = 1.0)
EXADJ	Adjustments to merchandise exports
EXMAZ	Merchandise export: maize
EXMIN	Exports of noncopper minerals
EXOG	Exports of miscellaneous goods
EXS	Export of services
EXTOB	Merchandise exports: tobacco
FDOLL	Foreign exchange: U.S. dollars per kwacha
FPOUND	U.S. dollars per pound sterling, market rate
IGMQNC65	Gross investment in noncopper mining and quarrying
INDTAX	Indirect taxes
LEMAQNR	Employment of expatriates in noncopper mining and quarrying
LEMAQZR	Employment of Zambians in noncopper mining and quarrying
PCLME	London Metal Exchange mean cash price per long ton of electrolytic copper wirebars
POP	Population: midyear estimates
PPM	Producer price of maize
PXD01G	Unit-value index of exports to developing areas: SITC 0, 1
PXD24G	Unit-value index of exports to developing areas: SITC 2, 4
PXD3G	Unit-value index of exports to developing areas: SITC 3
PXD5G	Unit-value index of exports to developing areas: SITC 5
PXD68G	Unit-value index of exports to developing areas: SITC 6, 8

PXD7G	Unit-value index of exports to developing areas: SITC 7
PXNC	Unit-value index of noncopper exports
RDCA	Depreciation rate: commercial agriculture
RDCN	Depreciation rate: construction
RDMF	Depreciation rate: manufacturing
RDMQ	Depreciation rate: mining and quarrying
RDSR	Depreciation rate: services
RDTC	Depreciation rate: transportation and communications
RTCS	Implicit rate of sales and excise taxes
SHCA	Sectoral share of nonmining real investment: commercial agriculture
SHCN	Sectoral share of nonmining real investment: construction
SHMF	Sectoral share of nonmining real investment: manufacturing
SHTC	Sectoral share of nonmining real investment: transportation and communications
SU	Net subsidies
TIME	Time-trend variable
VCAARATE	Rate of write-off of capital expenditure for tax purposes
VCTAXRATE	Company tax rate
VGO	Grade of ore from open-pit mining
VGOCO	Gross operating capital in open-pit mining
VGOCU	Gross operating capital in underground mining
VGOREALADJ	Adjustments to capital stocks in copper mining
VGPADJ	Adjustments to reported gross profits to reconcile identity
VGU	Grade of ore from underground mining
VMTAXRATE	Mineral tax rate
VOGRADE	Average grade of ore in copper mining
VUTTC	Total unit transportation costs of copper
XMD	Import duties, less imputed bank charges
XMD65	Import duties, less imputed bank charges in constant kwacha
XMNCOP	Value added by noncopper-mining sector

Model Identities

Output, Employment, and Wages

POPR	= POP − POPU
XMINQ	= XGDCOP + XMNCOP

XMINQ65 = XMINQ/PXMINQ
XMANF = XMANF65 * PXMANF
XCONST65 = XCONST/PXCONST
XTRCOM65 = XTRCOM/PXTRCOM
XAFFCS = XAFFCS65 * PXAFFCS
XAFFS = XAFFS65 * PXAFFS
XAFF65 = XAFFCS65 + XAFFS65
XAFF = XAFFCS + XAFFS
XSER = XSER65 * PXSER
XGDMC = XMINQ + XMANF + XCONST + XTRCOM + XAFF + XSER + XMD
XGDMC65 = XMINQ65 + XMANF65 + XCONST65 + XTRCOM65 + XAFF65 + XSER65 + XMD65
PGDP = XGDMC / XGDMC65
PXAFF = XAFF / XAFF65
LEMAQZ = VTLE / 1000.0 + LEMAQZR
LEMAQN = VTEE/1000.0 + LEMAQNR
LEINDZ = LEMANZ + LEMAQZ + LECZ + LETACZ + LEAGZ + LESERZ
LEINDN = LEEXPR + LEMAQN
LEIND = LEINDZ + LEINDN
AVWAGE = TCL * (1.0 + YWO/YWC)/(VTEE + VTLE)
BETA = TCL * (1.0 + YWO/YWC)/VSXCOP
WAGEIND = AVWAGE/0.0018244
ZMCOST = 0.2 * WAGEIND * FDOLL/1.4 + 0.1 * PXD7G/90.48 + 0.2 * PXD24G/105.0 + 0.5 * PXD68G/92.56
LPRICE/COST = log(VZPRICE * FDOLL/ZMCOST)

Aggregate Demand

YWDRPC = (YW − CGIT + YNOEX)/POPU/PBCPC
YPROFDRPC = (YPROF − XAFFS − YNIEX)/POPU/PBCPC
UBC65 = UBCP * POPU
BCPC65 = UBC65 + XAFFS65
BCPC = BCPC65 * PBCPC
BCGC65 = BCGC/PBCGC
CGLIC65 = CGLIC/PIGFC
IGFCP65 = IGFCPNM65 + KCMIN65 + IGMQNC65
IGFCP = IGFCP65 * PIGFC
DCGRR = CGRR − CGRR(−1)
IGFC = IGFCP + CGDC

IGFC65 = IGFC/PIGFC
INCHS = INCHS65 * PGDP
FINCI = BCPC65 + BCGC65 + IGFC65
DEFINCI = FINCI − FINCI(−1)
FINCIN = BCPC + BCGC + IGFC

Foreign Sector

EXCOP = EXCPRICE * EXCOPQT/1000.0
EXGD = EXCOP + EXMIN + EXTOB + EXMAZ + EXOG
EXCOP65 = EXCOP/PXGC
EXGS = EXGD + EXS + EXADJ
EXGS65 = EXCOP65 + (EXGS − EXCOP)/PXNC
PXGS = EXGS/EXGS65
MCAPRES = NINTR(−1)/PMGS
M01G = M01G69 * PM01G
M24G = M24G69 * PM24G
MCH = MCH69 * PM59
M3G = M3G69 * PM3G
MMATE = MMATE69 * PMMATE
MMAG = MMAG69 * PMMAG
MMMA = MMMA69 * PMMMA
M25G = M24G + M3G + MCH
M25G69 = M24G69 + M3G69 + MCH69
MG69 = M01G69 + M25G69 + MMATE69 + MMAG69 + MMMA69
MGFOB = M01G + M25G + MMATE + MMAG + MMMA + MGADJ
PMG = MGFOB/MG69/0.887794
MG65 = MGFOB/PMB
MSS = MSS65 * PMSS
MGS = MGFOB + MSS
MGS65 = MSS65 + MG65
PMGS = MGS/MGS65
EXCH = FDOLL/FPOUND
VZPRICE = PCLME * 0.9842/EXCH

Sectoral Investment and Capital Stocks

IGMQ65 = KCMIN65 + IGMQN65
SHMQ = IGMQ65/IGFC65

KSMQ65	= (1.0 − RDMQ) * KSMQ65(−1) + IGMQ65
VGOREAL	= (1.0 − RDMQ) * VGOREAL(−1) + 64.629 * KCMIN/(PXD7G/FDOLL) + VGOREALADJ
IGMAN65	= SHMF * (IGFC65 − IGMQ65)
KSMAN65	= (1.0 − RDMF) * KCMAN65(−1) + IGMAN65
IGCON65	= SHCN * (IGFC65 − IGMQ65)
KSC65	= (1.0 − RDCN) * KSC65(−1) + IGCON65
IGTC65	= SHTC * (IGFC65 − IGMQ65)
KSTRCOM65	= (1.0 − RDTC) * KSTRCOM65(−1) + IGTC65
IGCA65	= SHCA * (IGFC65 − IGMQ65)
KSCA65	= (1.0 − RDCA) * KSCA65(−1) + IGCA65
SHSR	= 1.0 − SHCA − SHCN − SHMF − SHTC
IGSER65	= SHSR * (IGFC65 − IGMQ65)
KSS65	= (1.0 − RDSR) * KSS65(−1) + IGSER65
KS65	= KSCA65 + KSC65 + KSMAN65 + KSMQ65 + KSTRCOM65 + KSS65

Government Revenues and Expenditures

PROF	= YPROF − XAFFS − VGP
MGTT	= MGFOB − MMATE − MCH
CGRR	= CGCD + CGEST + CGIT + CGMR + CGRO
CGRE	= CGSS + CGGS + CGES + CGREO
CGCE	= CGDC + CGLIC
CGBS	= CGRR − CGRE

Financial-Flows Sector

NNGD	= NNGD(−1) + DNNGD
MSQM	= MSQM(−1) + DMSQM
YPROFR	= YPROF − XAFFS − CGMR − INDTAX
YNIEX	= YIEX − YIM
BPEXG	= EXGD + EXADJ
BPT	= BPEXG − MGFOB
BPC	= BPT − MSS − YNIEX + YNOEX + EXS
BPCA	= BPPCX + BPPCG + BPPCO
BPNECF	= BPC + BPCA + BPEO
NFORA	= NFORA(−1) + BPNECF

National Income and Gross National Product

YWEXPR	= (WGAEXPR * LEEXPR + WGAMQN * LEMAQN)/1000.0
YWC	= (WGAMQZ * LEMAQZ + WGAMZ * LEMANZ + WGACZ * LECZ + WGATCZ * LETACZ + WGASZ * LESERZ + WGAAFFZ * LEAGZ)/1000.0 + YWEXPR
YW	= YWC + YWO
ULCNM	= (YW − TLC * (1.0 + YWO/YWC))/(LEIND − LEMAQZ − LEMAQN)
YPROF	= XGDMC − YW − CFC − INDTAX + SU
YN	= YW + YPROF − YNIEX + INDTAX − SU
XGNP65	= XGDMC65 − YNIEX/PMGS

Appendix B
Historical Values of Variables

	1958	1959	1960	1961	1962	1963	1964	1965	1966	1967
BCGC	45.808	53.576	56.523	65.631	71.256	77.418	79.561	115.102	123.602	158.252
BCGC65							81.168	115.102	115.876	135.150
BCPC	172.608	178.296	197.518	200.166	215.956	212.229	244.985	352.721	422.750	498.955
BCPC65							258.225	352.721	370.844	429.196
BPC							75.600	61.600	52.700	8.400
BPCA							-75.600	-65.600	-43.300	-11.400
BPEO										-20.400
BPEXG								366.800	446.000	465.200
BPNECF									9.400	-23.400
BPPCG										0.700
BPPCO							-75.600	-65.600	-43.300	-26.400
BPPCX							0.0	0.0	0.0	14.300
BPT								156.060	199.880	158.850
CFC	26.740	28.612	30.484	32.623	33.960	35.029	37.971	62.304	69.123	82.359
CGBS									95.200	84.400
CGCD								14.500	16.600	17.500
CGCE							22.628	30.572	65.700	107.000
CGCE65									62.400	95.300
CGDC							13.252	20.660	49.800	73.300

	1968	1969	1970	1971	1972	1973	1974	1975	1976
BCGC	174.339	181.504	198.500	272.600	302.400	326.500	347.000	420.000	480.000
BCGC65	143.050	139.209	163.000	204.600	220.000	214.000	212.000	230.600	230.000
BCPC	527.356	523.205	519.000	529.300	554.900	574.200	736.100	821.000	857.000
BCPC65	420.193	405.882	404.000	377.000	377.000	365.000	429.500	434.000	374.000
BPC	-2.900	338.300	77.000	-176.500	-148.800	93.400	48.500	-393.000	-65.100
BPCA	45.000	-165.300	37.800	57.200	92.600	-41.800	1.700	359.200	144.400
BPEO	-36.500	-45.400	-1.000	-76.100	-51.400	-63.000	-40.800	-118.300	-79.100
BPEXG	534.000	852.600	673.200	497.200	543.200	733.500	898.200	523.100	751.900
BPNECF	5.600	127.600	113.800	-195.400	-107.600	-11.400	9.400	-151.600	-39.700
BPPCG	20.000	7.100	-2.300	19.900	13.300	137.000	36.800	84.800	26.000
BPPCO	5.700	-6.300	-4.400	25.800	15.700	1.300	-9.200	-3.700	-7.200
BPPCX	19.300	-166.100	44.500	11.500	63.600	-180.100	-25.900	278.100	125.600
BPT	208.820	540.800	332.490	97.920	140.730	386.630	391.560	-74.510	282.870
CFC	112.709	148.540	146.000	176.100	205.700	216.300	242.500	265.000	305.000
CGBS	96.900	181.400	176.300	-18.300	-17.200	15.100	243.800	-98.700	-122.900
CGCD	21.200	31.000	32.500	36.700	49.800	46.000	60.900	48.400	29.000
CGCE	193.300	156.300	169.300	168.300	160.400	153.100	171.200	245.400	156.600
CGCE65	149.900	116.700	119.800	117.200	107.400	99.800	91.600	113.600	64.500
CGDC	126.100	104.100	97.400	116.500	103.300	97.500	117.200	131.000	103.900

	1957	1958	1959	1960	1961	1962	1963	1964	1965	1966
CGES								1.000	1.000	34.500
CGEST								4.280	7.758	8.600
CGGS								1.000	1.000	50.200
CGIT								12.172	17.114	21.300
CGLIC								9.376	9.912	15.900
CGMR								24.262	92.340	141.600
CGMRADJ								-79.088	-28.002	-76.132
CGRE										122.200
CGREO										7.600
CGRO								23.004	25.362	29.300
CGRPMT										2.600
CGRR								63.718	157.074	217.400
CGSS										29.900
CUQCZM	435.700	400.100	543.300	576.100	575.000	562.500	588.000	632.300	696.000	623.000
CUR	0.910	0.786	0.979	1.000	0.961	0.905	0.911	0.943	1.000	0.888
EXADJ									-8.200	-44.400
EXCH	0.500	0.500	0.500	0.500	0.500	0.500	0.500	0.500	0.500	0.500
EXCOP	165.000	135.400	214.400	239.200	222.800	221.800	238.200	296.800	343.200	460.600
EXCOPQT	415.200	400.600	522.600	556.700	553.800	544.900	587.400	681.700	683.500	599.200
EXCOP65									343.200	301.000

	1967	1968	1969	1970	1971	1972	1973	1974	1975	1976
CGES	63.700	62.700	69.100	79.800	102.400	81.900	113.100	101.200	158.900	120.500
CGEST	15.800	27.000	30.000	35.600	38.100	48.700	63.100	86.100	139.900	183.100
CGGS	65.300	86.900	80.200	95.300	128.300	137.500	135.100	164.000	238.200	207.000
CGIT	32.900	44.100	64.300	70.000	80.100	91.200	116.000	116.800	142.600	159.100
CGLIC	33.700	67.200	52.200	71.900	51.800	57.100	55.600	54.000	114.600	52.700
CGMR	163.900	176.100	235.200	251.100	114.100	55.700	107.600	328.800	59.400	12.000
CGMRADJ	-1.977	-18.273	-105.224	10.583	50.437	-9.865	-77.485	-11.202	59.400	12.000
CGRE	182.300	209.200	219.800	256.100	327.300	328.300	370.100	405.800	546.900	566.000
CGREO	9.900	7.700	13.100	15.200	17.300	20.700	26.800	36.300	36.900	115.800
CGRO	36.600	37.700	40.700	43.200	40.000	65.700	52.500	57.000	57.900	59.900
CGRPMT	4.100	2.600	8.800	8.800	25.800	14.100	10.300	9.000	13.700	2.700
CGRR	266.700	306.100	401.200	432.400	309.000	311.100	385.200	649.600	448.200	443.100
CGSS	43.400	51.900	57.400	65.800	79.300	88.200	95.100	104.300	112.900	122.700
CUQCZM	663.000	685.000	720.000	684.000	651.000	718.000	707.000	698.000	677.000	709.000
CUR	0.937	0.959	1.000	0.942	0.889	0.972	0.949	0.929	0.894	0.928
EXADJ	-1.800	-6.700	98.100	-37.300	17.200	7.200	-4.600	-2.200	5.100	3.283
EXCH	0.506	0.583	0.583	0.583	0.575	0.560	0.628	0.664	0.699	0.776
EXCOP	434.000	516.100	724.500	681.400	450.200	490.900	698.300	838.500	472.000	688.600
EXCOPQT	600.000	641.200	725.500	682.200	631.000	710.500	668.500	668.300	635.100	733.500
EXCOP65	302.000	323.100	366.800	343.700	319.100	357.300	336.700	338.400	322.200	374.700

	1957	1958	1959	1960	1961	1962	1963	1964	1965	1966
EXCPRICE	397.399	337.993	410.256	420.675	402.311	407.047	405.516	435.382	502.121	768.692
EXG		135.800	187.000	206.200	206.800	209.500	223.000	335.550	380.290	493.560
EXGD									375.000	490.400
EXGS									373.300	455.700
EXGS65									373.300	319.000
EXMAZ									1.900	1.800
EXMIN									17.300	17.800
EXOG									7.700	5.700
EXS									6.500	9.700
EXTOR									4.900	4.500
FDOLL	1.400	1.400	1.400	1.400	1.400	1.400	1.400	1.400	1.400	1.400
FPOUND	2.800	2.800	2.800	2.800	2.800	2.800	2.800	2.800	2.800	2.800
IGCA65									10.176	10.924
IGCON65									9.569	8.241
IGFC		93.261	89.658	80.543	84.571	77.788	69.946	80.755	151.885	219.704
IGFCP								67.503	131.225	169.904
IGFCPNM65								43.958	97.225	102.851
IGFC65								87.644	151.885	191.655
IGMAN65									10.480	15.524
IGMQNC65									12.629	-0.706

	1967	1968	1969	1970	1971	1972	1973	1974	1975	1976
EXCPRICE	723.333	804.897	998.622	998.827	713.470	690.922	1044.576	1254.676	743.190	938.787
EXG	470.010	544.420	766.490	714.960	485.180	541.560	741.960	905.090	521.050	752.050
EXGD	467.000	540.700	754.500	710.500	480.000	536.000	738.100	900.400	518.000	748.617
EXGS	475.200	544.500	862.700	685.400	500.600	586.100	780.500	944.500	575.000	756.000
EXGS65	322.500	346.000	434.500	404.000	360.000	410.000	386.500	400.500	415.000	460.000
EXMAZ	8.700	2.800	0.400	0.0	0.200	0.100	2.600	7.600	1.400	0.500
EXMIN	16.900	15.500	23.000	22.200	20.200	30.600	27.000	40.300	33.100	47.000
EXOG	3.700	3.600	3.400	4.000	5.900	11.700	5.400	8.200	6.500	7.417
EXS	10.000	10.500	10.100	12.200	3.400	42.900	47.000	46.300	51.900	51.000
EXTOR	3.700	2.700	3.200	2.900	3.500	2.700	4.800	5.800	5.000	5.100
FDOLL	1.400	1.400	1.400	1.400	1.400	1.400	1.541	1.554	1.554	1.402
FPOUND	2.767	2.400	2.400	2.400	2.434	2.502	2.452	2.339	2.222	1.806
IGCA65	15.385	17.710	17.428	12.864	18.250	13.386	13.502	12.388	14.072	11.410
IGCON65	13.987	14.359	14.523	20.368	9.786	8.439	11.762	10.792	12.259	9.940
IGFC	268.571	312.575	303.658	358.100	379.800	434.700	381.100	426.000	560.000	510.000
IGFCP	195.271	186.475	199.558	260.700	263.300	331.400	283.600	308.800	429.000	406.100
IGFCPNM65	123.573	96.068	136.021	177.594	124.450	160.398	122.559	113.917	164.373	137.587
IGFC65	233.109	239.319	223.436	268.000	264.500	291.000	248.500	228.000	259.000	210.000
IGMAN65	17.483	36.855	20.333	32.696	16.928	36.375	25.761	23.636	26.849	21.770
IGMQNC65	10.031	5.468	62.247	61.816	20.962	33.998	14.338	19.020	45.905	35.189

	1958	1959	1960	1961	1962	1963	1964	1965	1966	1967
IGMQ65								46.629	44.655	55.946
IGSER65								63.640	73.020	96.507
IGTC65								11.391	39.289	33.801
INCHS	21.400	-0.600	25.200	20.600	19.600	10.800	-16.200	37.300	52.200	50.700
INDTAX								105.477	160.056	153.208
KCMIN							27.000	34.000	52.000	52.900
KCMINREAL							27.802	34.000	50.591	49.345
KCMIN65							29.303	34.000	45.361	45.915
KSCA65							27.000	34.476	41.901	53.117
KSC65							15.200	20.669	25.448	34.999
KSMAN65							26.200	33.080	45.925	59.233
KSMQ65							522.100	541.628	560.774	588.352
KSS65							548.900	604.439	667.562	752.987
KSTRCOM65							130.400	138.291	172.037	198.882
KS65							1269.800	1372.583	1513.646	1687.571
LEAGZ			37.000	37.800	38.700	36.000	34.500	32.070	33.380	36.690
LECZ			36.000	30.000	23.900	23.400	29.100	43.100	60.310	62.700
LEEXPR			24.970	24.480	24.970	24.270	23.410	25.340	23.400	22.840
LEIND			241.200	234.700	228.100	221.500	233.300	261.060	290.250	309.080
LEINDN			33.000	32.600	33.300	32.500	31.700	32.880	30.680	29.370

	1968	1969	1970	1971	1972	1973	1974	1975	1976
IGMQ65	52.172	73.064	79.328	79.879	95.448	76.703	70.376	79.944	64.820
IGSER65	103.625	86.917	104.788	106.858	108.543	96.749	88.768	100.837	81.760
IGTC65	14.598	10.948	17.956	32.798	29.100	24.104	22.116	25.123	20.370
INCHS	57.800	-37.600	-11.500	47.400	21.300	82.700	203.000	30.000	-105.000
INDTAX	211.568	289.244	203.800	120.100	129.000	274.800	347.400	143.000	225.000
KCMIN	61.000	14.700	23.400	84.600	91.795	95.644	95.955	73.598	71.960
KCMINREAL	62.015	15.288	21.172	67.740	65.394	64.366	58.410	36.777	30.327
KCMIN65	46.704	10.816	17.512	58.917	61.450	62.366	51.356	34.039	29.631
KSCA65	65.547	76.378	81.561	91.677	95.927	99.836	102.241	106.089	106.890
KSC65	41.652	47.158	57.405	57.660	60.897	65.047	67.708	71.504	72.506
KSMAN65	91.126	104.957	129.218	135.521	162.216	175.648	185.935	198.653	205.326
KSMQ65	602.897	627.487	663.622	690.664	720.846	734.836	741.281	756.734	755.718
KSS65	845.131	916.265	994.844	1071.179	1164.755	1245.196	1316.531	1398.937	1461.111
KSTRCOM65	207.489	211.393	220.004	243.205	267.584	282.055	294.017	308.555	317.817
KS65	1853.842	1983.637	2146.654	2289.906	2472.225	2602.619	2707.713	2840.471	2919.367
LEAGZ	35.790	36.350	34.090	38.740	29.640	30.330	32.160	34.790	32.530
LECZ	64.400	59.690	66.220	63.140	68.230	66.630	66.270	67.790	51.340
LEEXPR	23.110	22.260	21.880	21.220	22.960	22.070	22.460	22.010	20.120
LEIND	319.200	328.290	342.970	365.550	367.930	373.440	384.890	393.490	368.360
LEINDN	29.050	27.820	27.390	26.550	34.140	33.390	33.300	32.320	32.500

	1960	1961	1962	1963	1964	1965	1966	1967	1968	1969
LEINDZ	208.200	202.100	194.800	189.000	201.600	228.180	259.570	279.710	290.150	300.470
LEMANZ	17.900	17.500	17.400	15.900	18.000	24.030	28.210	29.980	32.000	31.950
LEMAQZ	42.700	42.100	41.100	40.800	42.500	44.820	46.490	48.400	48.900	50.290
LEMAQN	8.030	8.120	8.330	8.230	8.290	7.540	7.280	6.530	5.940	5.560
LEMAQZR	2.862	3.261	3.967	3.232	3.297	4.429	4.015	5.464	5.911	6.650
LEMAQNR	0.524	0.403	0.630	0.616	0.964	0.631	1.494	1.408	1.252	1.161
LESERZ	64.200	64.400	64.100	64.600	68.800	72.290	73.510	83.510	89.200	102.560
LETACZ	10.400	10.300	9.600	8.300	8.700	11.870	17.670	18.430	19.860	19.630
MCH69					17.252	20.397	20.710	21.640	22.050	22.560
MGADJ					3.586	0.394	0.370	3.720	4.150	2.240
MGFOR					156.440	210.740	246.120	306.350	325.180	311.800
MGS						262.700	355.400	416.300	470.400	425.900
MG65					157.169	210.740	231.457	275.713	281.475	274.825
MGS65						262.700	327.000	391.500	420.000	390.000
MMAG					34.444	49.750	55.120	65.510	74.120	62.790
MMATE					42.420	69.590	97.940	126.330	134.440	123.040
MMMA					21.120	25.958	23.820	28.390	23.960	25.610
MSQM						107.600	145.300	169.400	218.700	281.500
MSS						51.960	109.280	109.950	145.220	114.100
MSS65						51.960	95.543	115.787	138.525	115.175
M01G					17.154	19.336	22.810	23.500	26.300	32.610

	1970	1971	1972	1973	1974	1975	1976
LEINDZ	315.580	339.000	333.790	340.050	351.590	361.170	335.860
LEMANZ	35.230	39.020	40.020	40.460	40.970	41.230	39.780
LEMAQZ	52.130	52.800	49.470	50.420	54.270	54.440	53.500
LEMAQN	5.510	5.330	11.180	11.320	10.840	10.310	12.380
LEMAQZR	8.036	7.803	3.225	2.133	2.534	1.448	0.418
LEMAQNR	1.135	0.579	6.580	6.815	6.448	5.815	8.320
LESERZ	108.190	124.290	123.190	129.740	137.390	142.430	140.510
LETACZ	19.720	21.010	23.240	22.470	20.530	20.490	18.200
MCH69	22.860	31.070	31.470	30.340	31.830	40.230	1.000
MGADJ	1.060	0.920	1.360	1.760	0.800	3.240	0.850
MGFOR	340.710	399.280	402.470	346.870	506.640	597.610	469.030
MGS	470.500	525.700	564.700	529.000	752.500	844.000	705.000
MG65	276.841	308.393	293.629	225.553	257.016	248.866	175.077
MGS65	394.700	418.200	425.000	353.000	385.500	360.000	231.000
MMAG	74.800	94.790	87.920	77.340	130.000	140.210	96.840
MMATE	131.720	160.120	168.010	138.910	167.800	211.300	167.040
MMMA	30.540	27.760	35.380	25.380	36.430	28.700	19.010
MSQM	355.600	324.900	368.600	473.400	489.700	532.800	582.500
MSS	129.790	126.420	162.230	182.130	245.860	246.390	235.970
MSS65	117.859	109.807	131.371	127.447	128.484	111.134	1.000
M01G	31.630	49.610	38.390	25.310	44.930	36.770	26.560

	1957	1958	1959	1960	1961	1962	1963	1964	1965	1966
M24G								3.950	4.960	7.220
M3G								17.446	20.600	19.600
NFORA									148.600	158.000
NINTR									142.600	150.400
NNGD									-88.700	-86.100
PBCGC								0.980	1.000	1.067
PBCPC		0.828	0.829	0.851	0.847	0.880	0.875	0.949	1.000	1.140
PC	44.360	45.323	46.217	46.500	46.600	47.200	47.000	48.400	52.300	57.700
PCLME	219.000	198.000	238.000	246.000	230.000	234.000	234.000	352.000	462.000	541.000
PGDP									1.000	1.245
PGDPNM									1.000	1.076
PIGFC								0.921	1.000	1.146
PMG								0.995	1.000	1.063
PMGS									1.000	1.087
PMMAG								0.918	0.968	0.906
PMMATE								0.828	0.764	0.983
PMMMA								0.773	1.021	1.000
PMSS									1.000	1.144
PM01G								0.904	0.892	0.949

	1967	1968	1969	1970	1971	1972	1973	1974	1975	1976
M24G	6.770	6.350	7.370	9.740	12.150	11.850	9.740	17.130	18.980	17.870
M3G	31.230	33.210	35.580	35.180	32.240	26.520	33.290	61.100	81.120	72.620
NFORA	134.600	140.200	267.800	381.600	186.200	78.600	67.200	76.600	-75.000	-114.700
NINTR	128.800	142.400	263.400	367.000	200.200	117.700	123.800	131.800	96.000	79.300
NNGD	-68.400	-37.300	-118.500	-163.800	18.600	147.200	205.000	78.100	317.800	571.100
PBCGC	1.171	1.219	1.304	1.218	1.332	1.375	1.526	1.637	1.821	2.087
PBCPC	1.163	1.255	1.289	1.285	1.404	1.472	1.573	1.714	1.892	2.291
PC	60.600	67.100	68.800	70.600	74.800	78.800	83.800	90.900	100.000	118.700
PCLME	411.000	517.000	611.000	589.000	444.000	428.000	727.000	878.000	557.000	781.000
PGDP	1.341	1.451	1.724	1.513	1.429	1.475	1.750	1.953	1.608	1.802
PGDPNM	1.198	1.281	1.343	1.324	1.406	1.467	1.539	1.707	1.850	2.082
PIGFC	1.152	1.306	1.359	1.336	1.436	1.494	1.534	1.868	2.162	2.429
PMG	1.111	1.155	1.135	1.231	1.295	1.371	1.538	1.971	2.401	2.679
PMGS	1.063	1.120	1.092	1.192	1.257	1.329	1.499	1.952	2.344	3.052
PMMAG	0.924	0.968	1.000	1.137	1.206	1.276	1.496	1.825	2.216	
PMMATE	1.010	1.072	1.000	1.125	1.166	1.262	1.477	1.535	1.963	
PMMMA	0.999	0.873	1.000	1.012	1.176	1.285	1.351	1.458	1.880	
PMSS	0.950	1.048	0.991	1.101	1.151	1.235	1.429	1.914	2.217	
PM01G	0.874	0.986	1.000	0.983	1.111	1.048	1.123	2.008	1.532	

	1958	1959	1960	1961	1962	1963	1964	1965	1966	1967
PXMANF								1.000	1.200	1.289
PXMINQ								1.000	1.563	1.655
PXNC								1.000	-0.272	2.010
PXSER								1.000	1.015	1.182
PXTRCOM								1.000	1.106	1.176
RDCA								0.100	0.101	0.100
RDCN								0.270	0.168	0.174
RDMF								0.137	0.081	0.091
RDMQ								0.052	0.047	0.051
RDSR								0.015	0.016	0.017
RDTC								0.027	0.040	0.040
SDISC	-1.800	4.800	0.400	-5.600	-1.800	-4.200	-9.600	0.0	0.0	0.0
SDISC65							-6.100	0.700	24.800	2.600
SHCA								0.086	0.075	0.082
SHCN								0.081	0.056	0.075
SHMF								0.089	0.106	0.093
SHMQ							0.334	0.307	0.233	0.240
SHSR								0.647	0.494	0.569
SHTC								0.097	0.269	0.181
SU								0.600	18.100	8.100

	1968	1969	1970	1971	1972	1973	1974	1975	1976
PXMANF	1.455	1.535	1.549	1.654	1.805	1.857	2.129	2.484	2.977
PXMINQ	1.859	2.508	2.011	1.498	1.499	2.397	2.739	0.738	0.881
PXNC	1.240	2.041	0.066	1.232	1.806	1.651	1.707	1.110	0.790
PXSER	1.232	1.310	1.285	1.327	1.384	1.490	1.622	1.720	1.910
PXTRCOM	1.201	1.242	1.254	1.320	1.444	1.561	1.736	1.962	2.205
RDCA	0.099	0.101	0.101	0.100	0.100	0.100	0.100	0.100	0.100
RDCN	0.220	0.216	0.215	0.166	0.090	0.125	0.125	0.125	0.125
RDMF	0.084	0.071	0.080	0.082	0.071	0.076	0.076	0.076	0.076
RDMQ	0.064	0.080	0.069	0.080	0.094	0.087	0.087	0.087	0.087
RDSR	0.015	0.019	0.029	0.031	0.014	0.014	0.014	0.014	0.014
RDTC	0.030	0.034	0.044	0.044	0.019	0.036	0.036	0.036	0.036
SDISC	0.0	0.0	0.0	0.0	0.0	0.0	0.0	0.0	0.0
SDISC65	16.700	14.600	2.200	20.400	16.000	6.400	-40.200	-25.500	5.000
SHCA	0.092	0.082	0.051	0.089	0.058	0.073	0.070	0.063	0.054
SHCN	0.075	0.068	0.081	0.048	0.037	0.063	0.061	0.054	0.055
SHMF	0.191	0.096	0.131	0.082	0.158	0.138	0.134	0.119	0.121
SHMQ	0.218	0.327	0.296	0.302	0.328	0.309	0.309	0.309	0.141
SHSR	0.566	0.703	0.665	0.622	0.620	0.596	0.610	0.652	0.648
SHTC	0.076	0.051	0.072	0.160	0.127	0.130	0.125	0.112	0.113
SU	12.000	16.600	18.900	27.700	25.300	35.000	47.400	69.000	60.000

	1957	1958	1959	1960	1961	1962	1963	1964	1965	1966
PM24G								0.913	0.888	0.895
PM3G								0.855	0.944	0.830
PM5G								0.946	0.988	0.929
POP		3.040	3.120	3.210	3.330	3.400	3.500	3.610	3.710	3.830
POPR							2.785	2.831	2.862	2.907
POPU							0.715	0.779	0.848	0.923
PPM							3.450	3.450	3.050	3.580
PXAFF									1.000	1.072
PXAFFCS									1.000	1.115
PXAFFS									1.000	1.057
PXCONST									1.000	1.223
PXD01G	100.910	97.030	96.040	94.080	95.060	96.040	98.000	100.940	98.980	99.000
PXD24G	115.790	105.600	101.000	101.000	103.000	99.000	100.000	101.000	105.000	103.000
PXD3G	114.730	104.300	99.910	97.000	98.940	97.000	97.000	97.000	97.970	94.000
PXD5G	123.130	119.540	114.400	113.360	110.240	106.080	104.000	104.000	107.120	105.000
PXD6RG	89.880	88.119	86.330	89.000	89.000	88.110	89.000	90.780	92.560	94.000
PXD7G	78.792	82.075	82.650	84.390	85.260	86.130	87.000	87.870	90.480	93.000
PXG		0.660	0.819	0.855	0.804	0.813	0.814	0.868	1.000	1.489
PXGC		0.659	0.816	0.854	0.804	0.810	0.813	0.867	1.000	1.530
PXGS									1.000	1.429

	1967	1968	1969	1970	1971	1972	1973	1974	1975	1976
PM24G	0.947	0.889	1.000	1.073	1.164	1.227	1.267	2.128	2.306	1.000
PM3G	1.024	1.046	1.000	1.015	1.076	1.132	1.150	2.814	3.742	1.000
PM5G	0.966	1.027	1.000	1.139	1.020	1.050	1.158	1.522	1.921	1.000
POP	3.900	4.010	4.060	4.180	4.300	4.420	4.640	4.750	4.980	5.180
POPR	2.894	2.915	2.868	2.907	2.940	2.968	3.089	3.094	3.207	3.281
POPU	1.006	1.095	1.192	1.273	1.360	1.452	1.551	1.656	1.773	1.899
PPM	3.510	3.690	3.690	3.790	4.270	4.260	4.290	5.000	6.300	7.125
PXAFF	1.112	1.146	1.166	1.176	1.303	1.329	1.418	1.485	1.440	1.739
PXAFFCS	1.197	1.130	1.146	1.231	1.472	1.435	1.416	1.353	1.074	1.473
PXAFFS	1.085	1.151	1.173	1.154	1.229	1.270	1.419	1.553	1.651	1.908
PXCONST	1.482	1.816	1.692	1.754	1.939	1.813	1.759	2.036	2.254	2.496
PXD01G	100.000	97.000	99.000	100.000	105.000	119.000	167.000	232.000	237.000	211.000
PXD24G	101.000	91.000	92.000	100.000	102.000	104.000	130.000	200.000	193.000	192.000
PXD3G	100.000	92.000	91.000	100.000	109.000	108.000	133.000	247.000	309.000	313.000
PXD5G	102.000	94.000	97.000	100.000	103.000	107.000	140.000	215.000	221.000	198.000
PXD68G	93.000	89.000	93.000	100.000	102.000	111.000	134.000	177.000	198.000	191.000
PXD7G	97.000	89.000	87.000	100.000	113.000	127.000	148.000	165.000	201.000	215.000
PXG	1.405	1.553	1.915	1.918	1.385	1.363	2.031	2.444	1.499	1.858
PXGC	1.437	1.597	1.975	1.983	1.411	1.374	2.074	2.478	1.465	1.838
PXGS	1.473	1.574	1.986	1.697	1.391	1.430	2.019	2.358	1.386	1.643

	1957	1958	1959	1960	1961	1962	1963	1964	1965	1966
TIME	7.000	8.000	9.000	10.000	11.000	12.000	13.000	14.000	15.000	16.000
TLC	41.988	38.796	42.314	61.649	62.408	61.453	61.435	66.425	70.520	77.848
VCAARATE	0.400	0.400	0.400	0.400	0.400	0.400	0.400	0.400	0.400	0.400
VCTAXRATE	0.450	0.450	0.450	0.450	0.450	0.450	0.450	0.450	0.450	0.450
VGO	4.990	4.870	5.120	5.370	5.090	5.520	5.450	5.240	5.400	4.220
VGOCO	45.482	48.412	50.056	50.198	52.518	54.774	55.352	58.814	63.598	63.048
VGOCU	17.950	20.522	23.960	27.522	34.860	40.306	41.114	41.596	44.342	45.648
VGOREAL	154.483	150.270	208.005	187.795	206.497	200.721	226.359	210.404	220.501	189.780
VGOREALADJ									-12.982	-65.697
VGP	94.336	66.842	145.194	163.652	162.540	136.410	140.238	180.544	190.200	271.172
VGPADJ	38.920	40.548	44.098	89.985	109.323	93.987	75.260	96.997	57.941	94.630
VGU	2.370	2.290	2.380	2.213	2.284	2.313	2.328	2.206	2.220	2.178
VIMNC	48.888	55.492	52.877	80.478	77.875	93.724	83.649	123.296	118.289	147.116
VKZPRI	387.900	330.600	412.150	429.780	403.330	409.940	409.940	435.520	526.270	705.850
VLORO	62.878	52.244	45.711	35.873	33.424	29.950	33.978	31.151	29.312	32.435
VLORU	116.748	106.521	107.906	89.061	91.157	87.609	91.473	79.964	80.507	82.015
VMTAXRATE	0.510	0.510	0.510	0.510	0.510	0.510	0.510	0.510	0.510	0.510
VOGRADE	3.105	3.084	3.247	3.228	3.236	3.445	3.382	3.232	2.990	2.860
VPCE	66.000	67.300	68.000	69.900	71.300	72.800	73.500	76.500	79.400	83.800

	1967	1968	1969	1970	1971	1972	1973	1974	1975	1976
TIME	17.000	18.000	19.000	20.000	21.000	22.000	23.000	24.000	25.000	26.000
TLC	95.730	89.298	97.577	99.664	105.454	97.103	105.718	117.117	108.816	105.780
VCAARATE	0.400	0.400	0.400	0.750	0.750	0.750	0.750	0.400	0.400	0.400
VCTAXRATE	0.450	0.450	0.450	0.450	0.450	0.450	0.450	0.450	0.450	0.450
VGO	4.010	3.950	4.000	3.650	3.430	3.520	3.730	3.630	3.400	3.430
VGOCO	69.124	74.598	84.538	227.636	256.256	305.304	356.822	419.304	469.232	499.373
VGOCU	46.084	47.080	48.840	140.451	166.685	197.357	232.631	248.182	165.673	164.659
VGOREAL	261.751	249.473	287.759	378.807	420.527	496.191	525.163	602.296	553.888	604.129
VGOREALADJ	35.656	-42.243	47.527	93.344	12.964	53.952	9.775	71.467	-30.047	68.800
VGP	256.802	285.790	501.158	346.800	150.600	158.600	325.100	503.819	-63.300	-37.946
VGPADJ	101.424	103.506	10.159	24.063	11.785	-16.492	-14.938	66.497	-43.661	-187.781
VGU	2.427	2.153	2.128	2.050	1.771	1.825	1.830	1.851	1.796	1.813
VIMNC	162.457	218.417	232.427	179.731	196.654	214.504	237.935	276.299	359.625	402.188
VKZPRI	751.200	827.710	1056.560	937.450	761.500	745.800	1089.900	1291.700	791.300	978.300
VLORO	45.855	43.685	40.156	61.197	62.712	58.821	60.099	68.751	70.331	69.525
VLORU	96.687	76.505	72.683	64.439	78.409	84.294	78.350	84.071	84.037	93.547
VMTAXRATE	0.510	0.510	0.510	0.510	0.510	0.510	0.510	0.510	0.510	0.510
VOGRADE	2.790	2.660	2.700	2.670	3.730	3.540	2.530	2.460	2.650	2.520
VPCE	88.200	95.600	100.000	105.000	110.900	118.700	126.400	138.100	149.900	173.600

	1957	1958	1959	1960	1961	1962	1963	1964	1965	1966
VQCZM	401.470	393.972	440.696	553.640	542.600	552.156	521.306	617.755	649.404	673.648
VSXCOP	161.214	135.300	214.677	239.460	219.965	217.617	235.474	296.802	349.572	429.215
VTEE	7223.000	6799.000	7492.000	7506.000	7717.000	7700.000	7614.000	7326.000	6909.000	5786.000
VTESP	415.608	409.256	520.871	557.168	545.372	530.852	574.410	681.488	664.244	608.082
VTLE	40806.000	35388.000	38881.000	39838.000	38839.000	37133.000	37568.000	39203.000	40391.000	42475.000
VTOPC	105.799	109.007	113.581	165.792	166.748	175.194	170.496	213.255	217.313	252.672
VTOTP	422.981	380.520	538.662	565.881	567.690	546.918	576.224	642.274	685.192	586.546
VTTCOST	14.923	14.719	18.390	23.665	26.466	20.017	25.411	23.533	28.504	27.708
VUMCO	153.424	161.600	150.700	136.449	146.272	151.710	124.491	110.621	119.118	126.676
VUMCU	86.360	120.847	104.030	147.466	135.786	182.235	178.791	242.044	203.743	264.889
VXCOP	97.404	65.089	143.410	135.316	115.624	103.876	126.413	149.972	202.778	254.390
VXNCCM	112.583	121.200	139.442	178.045	184.000	194.957	175.969	208.856	232.258	257.650
VXRCM	99.793	92.904	88.056	103.153	101.036	113.605	111.775	154.095	161.602	150.500
VXTO	112.583	121.200	139.442	178.045	184.000	194.957	175.969	208.856	232.258	257.650
VXTU	288.887	272.772	301.254	375.595	358.600	357.199	345.337	408.899	417.146	415.998
VZPRICE	431.083	389.746	468.482	484.230	452.735	460.609	460.609	692.882	909.407	1064.912
WGAAFFZ				120.000	120.000	128.000	132.000	176.000	172.000	190.000
WGACZ				236.000	240.000	254.000	276.000	286.000	322.000	322.000
WGAEXPR				2357.161	2392.086	2418.790	2506.039	2636.746	2903.567	3339.879
WGAMON	3813.200	3697.000	3390.400	5188.000	5178.000	5126.000	5128.000	5150.000	5378.000	6598.000

	1967	1968	1969	1970	1971	1972	1973	1974	1975	1976
VQCZM	542.454	648.368	697.259	686.365	636.101	698.122	682.990	709.450	647.840	711.681
VSXCOP	452.100	531.665	869.483	645.063	482.064	534.637	730.334	886.661	505.265	718.930
VTEE	5122.000	4688.000	4399.000	4375.000	4751.000	4600.000	4505.000	4392.000	4495.000	4060.000
VTESP	601.837	642.333	822.938	688.104	633.046	716.864	670.093	686.430	638.525	734.877
VTLE	42936.000	42989.000	43640.000	44094.000	44997.000	46245.000	48287.000	51736.000	52992.000	53082.000
VTOPC	296.722	349.381	378.484	322.326	343.249	359.545	390.296	449.339	524.903	569.095
VTOTP	518.861	659.747	755.351	686.365	636.101	700.564	682.990	709.840	647.840	711.681
VTTCOST	38.535	41.666	48.481	42.931	41.142	47.938	46.643	55.922	56.461	61.126
VUMCO	237.204	252.757	218.103	163.624	139.969	199.702	155.340	97.848	136.016	195.344
VUMCU	327.761	423.154	427.238	271.509	285.856	273.623	274.767	218.866	263.171	328.053
VXCOP	251.107	271.582	588.575	422.401	244.269	272.195	445.755	554.440	89.178	255.615
VXNCCM	187.132	201.533	216.405	397.648	404.498	438.347	440.007	408.753	408.666	385.414
VXRCM	118.392	161.243	173.940	306.629	250.355	242.734	278.788	281.121	288.564	280.994
VXTO	187.132	201.533	216.405	241.186	246.816	245.089	268.258	249.855	242.411	242.629
VXTU	355.322	446.835	480.854	445.179	389.285	453.033	414.732	459.595	405.429	469.052
VZPRICE	799.397	872.289	1030.886	993.768	759.859	752.810	1138.534	1300.565	783.732	990.273
WGAAFFZ	266.000	356.000	360.000	348.000	354.000	428.000	368.000	445.000	453.000	410.000
WGACZ	500.000	649.000	560.000	609.000	663.000	682.000	682.000	716.000	688.000	641.000
WGAEXPR	3570.261	3884.449	4417.137	4540.039	4883.711	4445.465	4837.008	5029.105	5004.184	4575.840
WGAMON	7608.000	7604.000	8174.000	7229.000	7336.000	5014.000	5406.000	6629.000	6784.000	7044.000

	1957	1958	1959	1960	1961	1962	1963	1964	1965	1966
WGAMQZ	354.000	386.000	435.000	570.000	578.000	592.000	596.000	732.000	826.000	934.000
WGAMZ				236.000	274.000	284.000	346.000	406.000	486.000	478.000
WGASZ				188.383	214.076	231.011	246.372	290.773	455.460	517.952
WGATCZ				292.000	345.000	388.000	426.000	482.000	486.000	688.000
XAFF									106.313	116.573
XAFFCS									25.696	31.313
XAFFCS65									25.696	28.083
XAFFS									80.617	85.260
XAFFS65									80.617	80.631
XAFF65									106.313	108.714
XCONST									61.232	82.042
XCONST65									61.232	67.071
XGDCOP		90.000	168.000	196.000	177.000	171.000	173.000	215.000	290.000	379.000
XGDMC		299.544	399.182	439.962	426.929	422.304	441.223	527.723	767.608	918.556
XGDMC65									767.608	738.074
XGNP65									722.108	681.574
XMANF									48.770	70.107
XMANF65									48.770	58.423
XMD									8.218	6.971
XMD65									8.218	11.406

	1967	1968	1969	1970	1971	1972	1973	1974	1975	1976
WGAMQZ	1322.000	1248.000	1412.000	1543.000	1569.000	1601.000	1685.000	1701.000	1478.000	1454.000
WGAMZ	668.000	644.000	744.000	802.000	946.000	1015.000	1046.000	1071.000	1179.000	1143.000
WGASZ	693.781	747.925	791.632	792.022	840.454	1127.047	1271.722	1223.411	1282.621	1221.639
WGATCZ	934.000	946.000	1034.000	1211.000	1495.000	1204.000	1292.000	1397.000	1834.000	1702.000
XAFF	119.957	124.591	129.344	136.100	154.000	171.700	186.300	191.000	195.500	247.000
XAFFCS	31.594	30.470	32.998	41.500	52.700	66.300	67.100	59.000	53.500	81.000
XAFFCS65	26.398	26.960	28.785	33.700	35.800	46.200	47.400	43.600	49.800	55.000
XAFFS	88.363	94.071	96.346	94.600	101.300	105.400	119.200	132.000	142.000	166.000
XAFFS65	81.443	81.754	82.112	82.000	82.400	83.000	84.000	85.000	86.000	87.000
XAFF65	107.841	108.714	110.897	115.700	118.200	129.200	131.400	128.600	135.800	142.000
XCONST	85.186	93.271	101.056	91.400	99.100	102.600	107.300	120.300	158.000	166.000
XCONST65	57.490	51.351	59.735	52.100	51.100	56.600	61.000	59.100	70.100	66.500
XGDCOP	379.000	411.000	637.000	457.000	297.000	318.000	536.000	626.000	145.000	195.000
XGDMC	1035.377	1146.170	1407.568	1278.999	1203.999	1334.699	1615.999	1904.000	1562.000	1793.000
XGDMC65	772.355	789.862	816.326	845.299	842.799	904.999	923.499	974.999	971.499	995.000
XGNP65	724.755	743.362	772.826	817.299	808.099	849.199	871.899	943.099	952.299	966.800
XMANF	87.482	107.498	115.728	127.500	142.600	182.300	196.800	249.500	279.000	320.000
XMANF65	67.872	73.867	75.391	82.300	86.200	101.000	106.000	117.200	112.300	107.500
XMD	8.502	10.655	15.756	15.700	16.300	18.299	2.600	15.400	26.000	18.000
XMD65	14.250	15.629	14.250	18.699	16.299	17.000	12.000	14.300	13.300	15.000

	1958	1959	1960	1961	1962	1963	1964	1965	1966	1967
XMINQ								305.870	398.638	398.847
XMINQ65								305.870	255.032	240.986
XSER								199.475	205.920	277.890
XSER65								199.475	202.805	235.030
XTRCOM								37.729	38.305	57.514
XTRCOM65								37.729	34.624	48.887
YIEX								62.500	76.600	78.400
YIM								17.000	18.600	27.800
YN	250.404	328.770	357.678	348.306	339.344	357.594	420.652	659.804	791.433	902.418
YNIEX	22.400	41.800	51.800	46.000	49.000	48.600	69.100	45.500	58.000	50.600
YNOEX								-3.500	-9.600	0.100
YPROF	103.228	183.810	211.504	193.603	184.172	194.244	235.764	330.088	392.559	422.311
YW	155.575	169.760	179.374	181.103	184.772	191.149	216.788	270.338	314.918	385.599
YWC			157.148	158.808	161.926	163.505	181.431	220.915	259.086	331.497
YWEXPR			100.518	109.603	103.097	103.025	104.420	114.127	126.186	131.225
YWO			22.226	22.294	22.846	27.644	35.357	49.424	55.832	54.102

	1968	1969	1970	1971	1972	1973	1974	1975	1976
XMINQ	432.600	670.126	467.700	303.700	326.100	544.700	635.000	156.000	204.500
XMINQ65	232.705	267.191	232.600	202.700	217.600	227.200	231.800	211.400	232.000
XSER	321.931	324.831	390.700	425.600	470.300	507.900	616.400	664.900	746.000
XSER65	261.240	248.028	304.100	320.800	339.700	340.800	380.000	386.500	390.500
XTRCOM	55.674	50.728	49.900	62.700	63.400	70.400	76.400	82.600	91.500
XTRCOM65	46.357	40.835	39.800	47.500	43.900	45.100	44.000	42.100	41.500
YIEX	59.500	63.800	62.200	65.900	88.000	87.700	81.400	50.900	89.000
YIM	7.400	16.300	28.800	22.300	13.900	10.400	19.200	5.900	3.000
YN	981.361	1211.528	1099.599	984.299	1054.899	1322.399	1599.299	1252.000	1402.000
YNIEX	52.100	47.500	33.400	43.600	74.100	77.300	62.200	45.000	86.000
YNOEX	-24.900	-51.000	-104.500	-107.800	-96.100	-80.800	-81.200	-79.000	-77.000
YPROF	410.915	553.127	465.400	365.800	388.800	487.700	583.500	368.000	383.000
YW	422.978	433.256	482.700	569.700	636.500	672.200	778.000	855.000	940.000
YWC	356.611	386.553	409.619	453.934	503.985	545.854	579.528	591.820	551.402
YWEXPR	134.937	143.774	139.168	142.733	158.123	167.947	184.811	180.087	179.270
YWO	66.367	46.703	73.081	115.766	132.515	126.346	198.472	263.180	388.598

Index

About the Authors

Chukwuma F. Obidegwu is currently with the Development Policy Staff of the World Bank. He was previously with the Institute of Policy Analysis, Toronto, and the Wharton Econometric Forecasting Associates Inc., Philadelphia, where the work for this book was conducted. He received the bachelors degree in mechanical engineering from the University of Nottingham, England, the M.B.A. from the University of Toronto, and the Ph.D. in finance from the Wharton School, University of Pennsylvania.

Mudziviri T. Nziramasanga is an associate professor of economics at Washington State University. From September 1977 to June 1979 he was a visiting professor and research scientist at the University of Michigan. He is currently on leave with the Ministry of Economic Planning and Development in Zimbabwe. His research interests have included commodity markets and economic development in Southern Africa, and he has been awarded grants by the U.S. Agency for International Development, the Ford Foundation, and the Carnegie Corporation's Commonwealth Program. Among his publications are *Nigeria: Dilemma of Nationhood,* with J. Okpaku (1969), and *Zimbabwe: Towards a New Order,* United Nations Conference on Trade and Development, 1980.

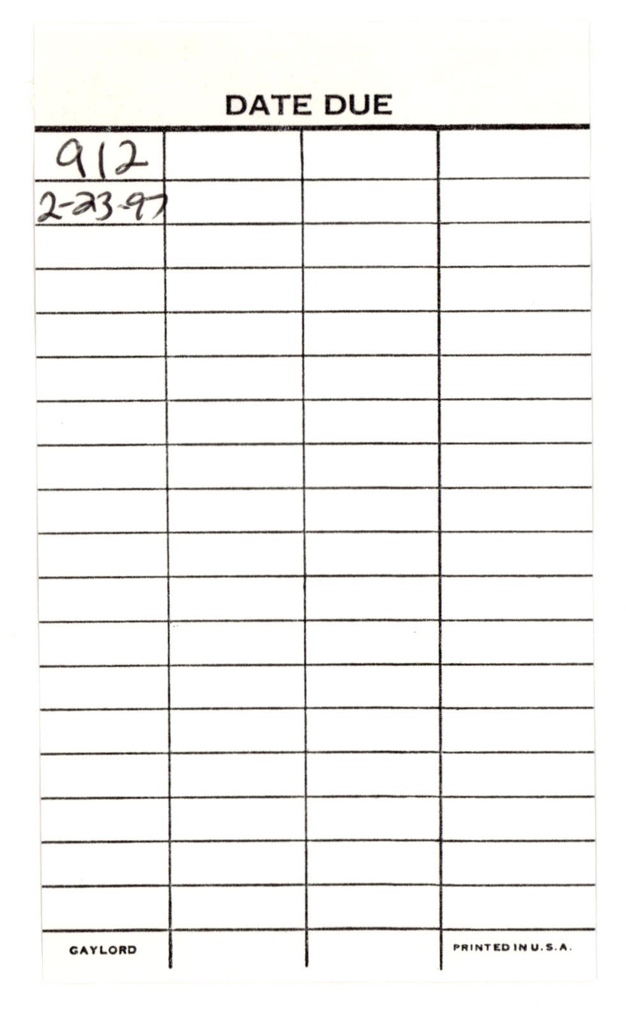